THE
SPINNING
MAGNET

The Force That Created the Modern World – and Could Destroy It

Alanna Mitchell

ONEWORLD

A Oneworld Book

First published in Great Britain, the Republic of Ireland and Australia by
Oneworld Publications, 2018

Published by arrangement with Dutton, an imprint of Penguin
Publishing Group, a division of Penguin Random House LLC

ISBN 978-1-78607-424-9
eISBN 978-1-78607-425-6

Typeset by Cassandra Garruzzo
Printed and bound in Great Britain by Clays Ltd, St Ives plc

Oneworld Publications
10 Bloomsbury Street
London WC1B 3SR
England

Stay up to date with the latest books,
special offers, and exclusive content from
Oneworld with our newsletter

Sign up on our website
oneworld-publications.com

MIX
Paper from
responsible sources
FSC
www.fsc.org   FSC® C018072

For James

contents

PART IV
switch

THE
SPINNING
MAGNET

playing with the universe

Night had set in by the time the green lights started dancing in the skies. I was in a canvas tent, trying to sleep in the bone-chilling cold of the late Arctic summer, worried about hungry polar bears. The cries of my fellow travelers roused me and I cursed, pulling on snow pants, boots, and a warm jacket before facing the night air.

The aurora borealis, or northern lights, were coursing through the black heavens, pulse after pulse of neon green against the scatter of stars, so near it was as if the curtain of light embraced us. They would fade away. We would hold our breath. And then back they would swoop, suffusing the sky. The heavens writhed with the otherworldly green rays, on and on, as if they held sway not just over our planet but over time too.

Watching those neon northern lights, I was closer than I knew to some of those who shared my impulse to understand the planet's magnetic force. My camping spot was on King William Island in the Canadian Arctic, about 100 miles or so from the Boothia Peninsula, where the British explorer James Clark Ross first pinpointed the Earth's magnetic north pole in 1831. His discovery of it was part of the magnetic crusade, the most sustained and impassioned scientific campaign the world had seen until then. At that time, the might of nations depended on naval prowess and efficient trade on the seas. And that depended on the magnetic compass. There was a trick to seafaring navigation, though. Knowing where you were depended on being able to adjust for the difference between magnetic north, where the compass pointed, and geographical north. The scientific world was united in an obsessive effort to figure out a formula that would allow sailors to know their coordinates more exactly. That required understanding the strange force that pulled the compass. And that demanded information from the top and bottom of the Earth, where the force showed itself most strongly.

King William Island itself resonates with a grisly piece of the magnetic quest. It is where the British explorer Sir John Franklin vanished in the 1840s along with his 128 men and their two ships. They were trying to complete the Northwest Passage, the quick sea route over the top of North America that would connect the wares of the Orient with the markets of Europe. But Franklin was also a player in the magnetic crusade. The ships in his Arctic adventure, HMS *Erebus* and HMS *Terror*, carried enough equipment to set up a state-of-the-art magnetic observatory in the Arctic, one of dozens that were being established all over the world as scientists tried to decode the secrets of the magnet. Franklin himself orchestrated the setting up of a magnetic observatory on the island we now know as Tasmania, off the Australian mainland's south coast.

But when his Arctic ships got stuck in the ice and Franklin died, along with many of his men, the survivors abandoned ship and took to the frozen island in their leather-soled shoes and navy cloth great-coats in a bid to walk to safety with the Earth's magnetic field as their trusted guide. Some resorted to cannibalism. All died. Few of their skeletons have been recovered. It was the worst disaster to hit Arctic exploration. The Inuit who live on King William Island say the sailors' ghosts haunt the place still. Among the relics recovered from the sailors' doomed march was a brass pocket compass, currently in the collection of the National Maritime Museum in Greenwich. The men were trying to read their magnetic position even in those grim final days. It was their last hope to get home.

Franklin, Ross, and other Victorian explorers, stuck for years on end in the Arctic, undoubtedly saw the northern lights. But they could not have known how the compass, the magnetic poles, and the auroras fit together. Today, we know that they are facets of one another. The Earth is a giant magnet with its own two poles, north and south. Stretchy magnetic field lines leave the surface of the Earth at the south magnetic pole; run around the planet, where they interact with the magnetic fields of the sun and the galaxy; and then reenter the Earth at the north magnetic pole in unending, erratic loops.

Our magnetic field is generated in the Earth's most secret inner reaches—its hot, yet frozen, metal inner core surrounded by a liquid metallic outer core. That heat, a remnant of the planet's violent birth, is the secret to the planet's magnetic power. The core has been on a multibillion-year quest to get rid of that heat from the inside out and is shedding it through convection. Convection generates electrical currents in the molten metal of the part of the core that is not yet solid, and those currents produce a magnetic field. Phoenix-like, that field is continually being created and destroyed. It stretches out for thousands of miles into space, our planet's giant defense system against the lethal invisible rays and charged subatomic bits that would otherwise rip through living tissue and tear away the Earth's atmosphere. Consider that our sister planet, Mars, lost its atmosphere, water, and likely any life forms when its internal magnetic field died billions of years ago.

A compass, with its magnetized needle, is responding to the Earth's magnetic field. And the auroras? To humans, the magnetic field is invisible and imperceptible. We see the *effects* of the Earth's field, for example, when a compass needle moves. But many species actually perceive the magnetic field. Some scientists call it a magnetic sixth sense that is like sight or touch, just more poorly understood and more complex. Creatures from bacteria to spiders to squids to sea turtles to almost everything with a backbone somehow use the magnetic field to navigate; it's one way they find food, mates, homes. Birds, though, are in a class of their own when it comes to perceiving magnetism. One study found that they can just open their eyes and see it, the way we see light. Biologists believe that humans once had the ability to sense the field the way other vertebrates still do. Vestiges of it are knitted into our genetic makeup, albeit dormant. But mostly we walk around unaware of this hidden force field that has such an effect on our lives and our world.

The exception is the auroras. They usually appear in great oval rings around the top and bottom of the planet. Occasionally they show up closer to the equator. They are magnetism briefly made visible, the product of violence in outer space perpetrated by plasma that has roared at us from the sun. That plasma, also known as solar wind, has a magnetic field of its own, and when it is directed in a certain way, it can tear open the Earth's magnetic field. Solar wind rushes in along the loops of the Earth's magnetic field, pouring fast-moving, highly energized atomic particles into the polar regions, where they crash into oxygen and nitrogen atoms in the Earth's upper atmosphere. In turn, the solar wind's ferocious energy is transmitted to the oxygen and nitrogen atoms, exciting them. As the atoms relax back to normal, they shed their extra energy as light and color. The green northern lights I saw were excited oxygen atoms frolicking in the sky, showcasing the Earth's own magnetic field. It was like looking into the sky to see a reflection of the machinations of the bowels of our planet.

For thousands of years, men and women have struggled to understand what magnets mean. They have looked to the heavens, not because they thought the aurora or the celestial bodies could provide clues about magnetism but because they thought that the heavens were the Earth's puppeteers. If they looked closely enough at the stars, they would be able to read whatever they needed to about our planet. Painstakingly, through experimentation and flashes of inspiration and, finally, maths and theoretical physics, they built up the conceptual understanding of magnetism we have today. It is highly abstract. It is ardently creative. It is slightly imperfect. But it is powerful.

And revelatory. It tells us the surprising news that we need to pay close attention to what used to be called our planet's magnetic soul. The Earth's magnetic power is on the move. That power is eccentric and, therefore, so are its poles. Eventually, the covert intrigues within the Earth will become so violently disruptive that they will force the poles to switch places. We know this because it has happened hundreds of times in the planet's history. The last time, 780,000 years ago, our species was not yet on the planet. But the long string of pole flips has left traces buried in the seams of the plates that fashion Earth's crust and in some of the rocks and lava laid down on top of it. When the poles switch again, the one we call north will move to the south. South will be north. As that happens, the magnetic swaddling that protects our planet will waste away to only about a tenth of its usual vigor. In turn, that will affect each of us and the very fabric of our civilization. As a side note, the auroras, normally seen only in higher latitudes, will likely be visible closer to the equator as solar wind tears more brutally through our atmosphere. When that happens, it will prophesy calamity.

Our Earth's magnetic field is being made even now in the core. In its turn, the Earth is continually being buffeted by the magnetic field the sun generates within itself, which is within the one our galaxy makes. Most of the planets in our solar system make their own magnetic fields. And they are all related to the universe's electromagnetic field, one of the fields of fluid-like substances that flow everywhere.

The fields can show up in specific places as particles, such as electrons and quarks, which in turn make atoms. As I was trying to understand all this, I spoke with the American theoretical physicist Sean Carroll from Caltech. Theoretical physicists are the poets of science, the ones who can see into the tiny guts of matter and imagine what happened to create the stuff of the universe. At what point in the birth of the universe did the fields and particles show up? I asked him. Fields and particles didn't show up, he replied. They were always there. Fields are what the universe is made of.

For example, when you try to push the north end of a bar magnet toward the south end of another bar magnet, they snap together. But try to put two norths together and it won't work, no matter how hard you push. It looks for all the world as if nothing is between the two magnets. But in fact that space is filled with the universe's immensely powerful electromagnetic force. It is there whether our planet is making its own magnetic field or not. When you play with magnets, you are playing with the universe.

The research for this book has also taken me more deeply into the world of chemistry. One entrée was through my son, Nick Michel, who is studying to be an organic chemist at the University of Toronto. He has patiently sat with me to explain how chemists understand the inner logic of atoms and molecules. I, a Latin scholar, have found myself reading the textbooks of chemistry, physics, and biology. My hunger to understand the planet's vast magnetic system has prompted me to travel to universities in several parts of Europe and North America, seeking explanations among some of the world's top scientists. They are the particle physicists who look into the bits that make up atoms and also the astrophysicists whose professional sphere is the machinery of other planets and the stars, including our own sun. They are the geophysicists who yearn to understand how our planet functions and how it once did, and who want to foretell our future. Many of them seemed to be as much at home hacking a piece of rock out of the Earth's crust as crunching complex numerical simulations through supercomputers.

They have pointed me back in time to the early metaphysicists, to the era when science, magic, and religion were the same thing. For much of the time humans have been investigating magnetism, it was a dangerous enterprise that threatened the theological ideas that dominated society. The very name of this book, *The Spinning Magnet*, would have been considered heretical at other points in history because it is a different way of describing the Earth from what the Bible says.

Modern scientists have helped me understand the magnetic investigations of the Middle Ages, the electrical exploits of the Renaissance, and the compulsions of the Victorians. Each of these generations of magnetic explorers had its own philosophies and developed novel explanations for what they found. Each tried to explain their findings in words, often in metaphors, sometimes inventing them for that purpose. To describe a magnet's poles, for example, is to invoke the metaphor of a planet spinning on its axis in the sky. It is the language of early astronomy, yet it persists today. Other concepts are explained in the language of watchmaking—clockwise and counterclockwise. Or direction—up and down. Or classical physics via astronomy—orbit and spin. It's a bit of a jumble. And the fact is that even today's primary way of describing the world in the language of quantum physics—orbitals and fields and superpositions—has not caught up with the ideas it is trying to explain, as Carroll points out.

What that means is that the understanding as well as the language is evolving over time, and will continue to evolve. It also means that the language of one branch of science—say, chemistry—does not necessarily speak the language of another—say, theoretical physics. What you are about to read is a translation of some of these ideas from science into journalism. I hope it throws its own kind of light.

PART I

magnet

I really can't do a good job, any job, explaining magnetic force in terms of something else that you're more familiar with because I don't understand it in terms of anything else that you're more familiar with.

—Richard Feynman, Nobel laureate, 1983

CHAPTER 1

the beginnings of things

Jacques Kornprobst, the man who can read the secrets of the rocks, was agitated. He had arrived twenty minutes early to pick me up at the hotel in Clermont-Ferrand, an ancient French university town perched on an annealed crack in the planet's crust. He had the entry code at the ready to get into the free parking lot behind the building. The code had failed him.

Some drivers cruise the streets nonchalantly, certain that the perfect parking spot will open up at just the right time. Kornprobst was not among them. Parking in this city of 150,000 had become troublesome over the decades he had lived there, and as he had mapped out the day's tightly choreographed itinerary he had made intricate plans about where to park. And now, the first parking spot of the day had fallen through.

Inside the hotel he sprinted, red-faced, fingertips frigid in the spring chill.

"Kornprobst!" he rapped out as he met me for the first time. Then he turned swiftly to the reception desk to let off a stream of injured French, explaining to the bewildered woman sitting there—she had been so friendly earlier, solicitous about replenishing the croissant basket and tinkering with the café-au-lait machine—about the affront. He had called the day before to secure the code. And now, today, he said, chin thrust slightly forward, it was malfunctioning.

Abruptly, she left through a back door. He darted out front to a tiny blue Renault car that was parked haphazardly on a curve at the corner, performed a roundabout U-turn through the city's tortured roads, and then nosed up to the gate with its uncooperative code. The receptionist stood there, punching in numbers, shivering. He drummed his fingers on the steering wheel. Finally, the barrier began to rise and the receptionist, without so much as a glance behind her, returned inside to her desk. Kornprobst smiled grimly, thrust the little car into gear, gunned the engine, and zoomed triumphantly into a parking spot.

He was watching the clock. He was on a mission to memorialize the life and work of Bernard Brunhes, a French physicist who, along with his research assistant Pierre David, made an astounding, violently unsettling, and controversial find at the turn of the last century. Brunhes, whose name is pronounced "brune," discovered that the planet's two magnetic poles—north and south—had once switched places. In the decades following his discovery, his colleagues, originally aghast at Brunhes's finding, proved that the poles have reversed not just once, but many times on an unpredictable, or "aperiodic," schedule. The last time was 780,000 years ago.

But despite the fact that our current magnetic epoch is named after him, Brunhes has largely slipped out of the scientific memory. He does not even rate his own entry in the *Encyclopedia of Geomagnetism and Paleomagnetism*, the bible of the discipline of reading patterns in the Earth's magnetic fields. Nor is he lionized in France, usually so careful to honor its own. In fact, he's all but unknown even in his homeland, along with his grand scientific finding that the poles can switch places, that up can become down.

Kornprobst, a fellow physicist, felt that he must right this wrong. He was so committed to Brunhes's memory that some years ago he took the trouble to find the spot in the countryside where Brunhes hacked a piece of crumbly terracotta rock—similar to the stuff of Greek vases—out of a roadcut and made his great discovery. Kornprobst painstakingly pieced together the clues about where it could be and is one of a handful of people in the world who can usually find it. The first time he made the pilgrimage to the site, he left frustrated, having failed to identify the right seam of rock. He's found it several times since, but it's so overgrown, so unmarked, that success is always touch and go.

Kornprobst thought that Brunhes should at least have a commemorative panel at the university in Clermont-Ferrand, so he sweated through a couple of years writing to geological agencies and eminent physicists all over the world jostling their elbows about Brunhes's contribution to science, raising the money to erect it. Then he arranged for a ceremony and lecture to accompany its inauguration at the university in 2014. It was through that ceremony that I found Kornprobst. He wrote an article about it for *Eos*, a journal of the American Geophysical Union. I read it and sent him an email asking him if he would help me understand why Brunhes was so important and maybe even find that seam of terracotta. He wrote back thirteen minutes later to say he would be delighted. I was at the hotel in Clermont-Ferrand two weeks later.

Sporting a thick, off-white cable-knit sweater the same hue as his rakish hair, Kornprobst left the car in the lot and we set off briskly on foot from the hotel through the back streets of Clermont-Ferrand. It is one of the oldest cities in France, founded more than two millennia ago on the site of what was then a sacred grove of trees. And so we were marching through time, across the history of science. Up the road named after Pierre Teilhard de Chardin, a Jesuit priest and paleontologist who deeply offended the Vatican for asserting that the

book of Genesis is more allegory than fact. Past the geology department of the downtown campus of Université Blaise Pascal, named
after the seventeenth-century mathematician and physicist whose
seminal experiment on barometric pressure was conducted a few kilometers outside the city by a brother-in-law ("There is the belief that
Pascal experimented with pressure here," Kornprobst declaimed,
pointing vigorously down the street, "but it's not true!"). Across a
road named after the nineteenth-century zoologist Karl Kessler. And
finally, to rue de Rabanesse, named after the tiny pale stone Renaissance castle that was Brunhes's home and first observatory.

Kornprobst gestured to it triumphantly, eyebrows raised, as if it
explained a great deal.

It looked like nothing out of the ordinary. It was standing forgotten on an overgrown patch of land across the street from a busy art
school, surrounded by two layers of forbidding wire fence. Many of
its lower windows—once elegant—were partially filled in with cement blocks. The plaster that had covered the volcanic fieldstone that
made up its walls had decayed, leaving gaps along the seams so you
could see how it had all been fitted together. Its turret, where Brunhes collected meteorological information beginning in 1900, was still
sturdy, reaching six floors into the sky, fifteenth-century iron fret-
work still robust.

This observatory is where the tale of Brunhes begins. And where
the tale of Brunhes begins, so too does the story of the discovery of
the planet's long string of pole reversals. And that story, in turn, contains the tale of the mysterious magnetic organism in the core of the
planet and how it has become deeply disturbed once more, yet again
deciding whether to reverse.

It was here that Brunhes, whose name means "brown" in the Oc-
citan language of the ancient troubadours of this land, began to dream
of understanding magnetism, the Earth's secret power. We never feel
it and rarely see it, but all the same, scientists and philosophers have

been trying to understand it for thousands of years. For most of that time, people have imagined it to be local and transient. Magic, even. And fickle magic at that. In fact, magnetism is one of the few essential powers of the universe. To understand it, you have to go back in time to the birth of the universe, to see how the universe is arranged. And you have to do that in the company of theoretical physicists, who have developed the most precise mathematical laws so far to describe reality.

CHAPTER 2

the unpaired spinning electron

Today, magnetism is properly known as electromagnetism, one of the universe's four fundamental physical forces. A fundamental force is one that simply exists. It is a never-ending characteristic. If you compare it to mathematics, it's conceptually akin to a prime number—like 3 or 13—that can't be divided into any combination of whole numbers except itself and 1. A fundamental force can't be reduced into a more basic force; it simply is.

In theory, there are an infinite number of prime numbers. But in the universe today, there are only four fundamental physical forces—at least that we know of: gravity, strong nuclear interactions, weak nuclear interactions, and electromagnetism. (Caveat: scientists continue to look for a mysterious fifth force and make occasional, highly contested claims that they have found it. Stay tuned.) Each of these forces is intrinsic to the workings of the universe, indispensable, inescapable. They were born along with the universe, the sun, stars, moon, and skies.

Gravity is the force that made Isaac Newton's apple fall to the ground and that keeps you from falling off the face of the Earth as it spins. It governs bulk matter and attracts but doesn't repel. It is the weakest of the forces but stretches to infinite space. The nuclear interactions govern the insides of atoms but nothing larger. Strong nuclear interactions hold the cores of atoms together. Weak ones (called weak because their sphere of influence is even smaller than strong nuclear interactions) allow atoms to fall apart and metamorphose into other types of atoms. That makes the weak nuclear force the ultimate alchemist. It is responsible for radioactive decay. The energy of our sun, which makes Earth the warm, livable place it is, is the result of both types of nuclear forces. As you read this, the weak interaction is allowing hydrogen protons to shed enough energy to become heavy hydrogen (deuterium) and then the strong interaction allows the atoms that result to fuse together into helium atoms.

So what is electromagnetism? It is the force that holds matter together. Apart from gravity, which holds us down on Earth, everything we see around us is due to magnetic and electric forces, explained the American theoretical physicist Sean Carroll. It is the basis of the structure of the atom, holding electrons in place and allowing atoms to link up into molecules. But where did the structure of the atom come from? From the birth of the universe itself.

So, Big Bang, about 13.7 billion years ago. The universe is created. What makes up the universe and everything in it? Is it atoms and the elements they form? To quantum field theorists, the answer can be stripped back to something more fundamental than atoms. To them, the universe is fashioned of fields: a field for each of the fundamental forces and thirteen other fields governing matter. A field is simply a mathematical way of talking about fluidlike substances that are spread out everywhere throughout the universe and have a value everywhere in the world. They ripple and sway. It's a difficult concept. In his famous physics lectures to undergraduates at the California Institute of Technology, the late American physicist Richard Feynman said he had never been able to develop a mental image of the electromagnetic field: "How do *I* imagine the electric and magnetic field? What do *I* actually see? What are the demands of scientific imagination? Is it any different from trying to imagine that the room is full of invisible angels? No, it is not like imagining invisible angels. It requires a much higher degree of imagination to understand the electromagnetic field than to understand invisible angels."

Some portions of the electromagnetic field can be discerned. A wave of light is a bump in the electromagnetic field that travels through space. A particle, on the other hand, exists in only one location and nowhere else. But, like light, a particle is still a facet of a field, a little wave tied up into a bundle of energy. And particles make up atoms, or the stuff we can see and feel. The most basic particles, for our purposes here, are electrons and two kinds of quarks: up and down. Each of them has its own field. If you were to think about it in biological terms, they are like the base pairs of DNA that are the foundation of every living thing on Earth. The magic of the universe is that, conceptually, any of these quarks could be exchanged for any other quark. The same goes for electrons. They and their fields are the building blocks of all matter, including you.

The inevitable implication of this, to a theoretical physicist, is that what we observe is only a portion of what is there. What we normally think of as empty space is filled with this powerful electromagnetic force field that gives matter its concreteness, as well as the other forces and fields. To physicists, this is humdrum reality.

By the time the universe is a few millionths of a second old, it has cooled down enough for quarks to join together to create protons and neutrons, the bits that will eventually form the cores of atoms. (The word "atom" comes from the Greek meaning "indivisible." Wrong, as it turns out.) Electrons don't join up to make anything bigger; they remain solo. These particles aren't forming atoms at this point; the universe is still too hot. They're just bits.

At about the 100-second mark in the life of this new universe, things have cooled down enough for some protons and neutrons to link up and make the heavy centers, or nuclei, of helium atoms—two protons, two neutrons. Give it another 380,000 years and now it's cool enough that some of those simple nuclei have got electrons in the space around them. The electrons are negative. The protons are positive. They are responding to the maxims of the electromagnetic field: Opposite charges attract and like charges repel. So the negative electrons are drawn to the positive protons. That attraction keeps the electrons inhabiting the space around the nucleus. Neutrons, as the name suggests, are neutral. Why are protons positive and electrons negative and neutrons neutral? No one has satisfactorily explained that; they seem simply to have been born with those differences and we happened to endow them with that nomenclature. Why do opposite charges attract? Again, it just seems to be part of how the fields showed up.

Most of the atom's weight is in its center, in the protons and neutrons that are the nucleus. The electrons are lightweights, usually in motion. Some chemists like to say that if the whole atom were the size of a baseball stadium, the nucleus would be about the size of a baseball in the middle. That means most of an atom is what the early theorists of atomic structure used to think of as empty space. Today we know that it is filled with invisible fields. Because atoms create matter, that also means that most matter, not just space, is invisible fields. That includes the matter that makes up your body. I sometimes imagine what it must have felt like for the scientist who figured that out. I imagine him looking at his hand with renewed intensity, trying to peer through it.

It's the arrangement of these three main components of an atom—electrons, protons, neutrons—that determines which type of atom is which. If you can wade through a few more points here, you'll get to one of the ideas that lies at the heart of magnetism.

The number of protons is key. That number determines which

element it is. In other words, an element's very identity is controlled by the number of protons in its nucleus. So is its order in the periodic table of elements, because the periodic table is arranged by ascending atomic number, from hydrogen on up.

When the number of protons changes—for instance, during radioactive decay or nuclear fusion—then the name of the atom changes as well. So hydrogen is hydrogen because it has only one proton in its nucleus. When immense heat forces a hydrogen nucleus to fuse with another hydrogen nucleus, the atom that emerges has two protons, and therefore it is helium. As goes the number of protons in a nucleus, so goes the name of the element.

By contrast, the number of neutrons and electrons in an atom can shift around without changing the atom's name. So carbon, for example, the sixth element on the periodic table, will always have six protons. But sometimes, in nature, it has different numbers of neutrons. Those variations are called isotopes. Too many neutrons, and an atom becomes radioactive and unstable and wants to metamorphose into a different, more stable element.

It's the electrons, which inhabit the space around the nucleus, that provide one of the secrets to the puzzle of electromagnetism. Just over a century ago, when electrons were discovered, scientists imagined them as little planets moving in a fixed track, or orbit, around a home star, or nucleus, just as the Earth does around the sun. They even used that imagery in the names they gave things, like orbits.

Today, they say instead that electrons move in orbitals, which are mathematical expressions of where electrons probably are. To me, it sounds like gobbledygook. But it just means that electrons are not in a track but somewhere in a pretty well-defined three-dimensional cloud around the nucleus. Probably. You can't point to a spot and say that's exactly where an electron is right now. And it's not necessarily in an orbit. Orbitals have many theoretical shapes, some spherical, others complicated three-dimensional figure eights or dumbbells, others far more complex.

The basic, extremely counterintuitive, but absolutely critical idea is that the electron and other particles operate as both a fluidlike component of a field and as a single physical particle *at the same time*. They are components of the fields that form the universe. For example, when electrons jump from one orbital to another, they are acting as individual physical entities. But when they are incapable of being in a single identifiable place at any time, they are acting like a wave or a field. To understand electromagnetism, we have to live with the complexity of this.

There's a little more. While the planetary language to describe electron behavior is now obsolete, it is still helpful as a mental image. So is the fact that the orbitals are arranged in groups of concentric rings or layers or shells around the nucleus. That simplification makes them a little easier to visualize. A central point is that the farther away from the nucleus the electron is, the more energy it has and the more apt it is to be able to be persuaded to move away from an atom's influence.

The idiosyncrasies of these electrons in orbitals provide one way to create a magnetic field. With some exceptions, every electron in the universe is held in one of these orbitals or is in the process of moving into one. But one of the unbreakable rules of the universe is that each orbital has room for only two electrons—a pair—and the electrons in a pair must spin in opposite directions. Confusingly, here the metaphor to describe direction comes from watchmaking: If one in the pair spins clockwise, then the other must spin counterclockwise. The point is that one movement must offset the other in order to reach balance. In addition, each orbital is contained within a shell, or grouping, that can hold a fixed number of pairs of electrons.

Electrons have strong preferences when it comes to inhabiting their orbitals. They fill them up in highly regimented ways. In fact, they have a rigid code of conduct that can be broken only in exceptional circumstances, filling up one shell of orbitals before moving on to another. Going back to the image of the baseball stadium, it's a little like filling up seats at ground level close to the diamond first, and then, if there are enough electrons, moving higher up and farther away, section by section, row by row.

Electrons strongly prefer not to be in pairs. They'd rather have a solo slot in an orbital. Pairing is a last resort. But they'll do it before spending the energy to move to a higher shell, or level in the stadium, where they could have a slot all to themselves.

A rather unselfconscious university chemistry teacher, perhaps remembering a beer-soaked baseball game from his youth, once described it this way: Imagine you have six young men who urgently need to urinate, and only three urinals. Each of the first three men in the line will naturally take a separate urinal. At that point, each of the others will ask to share one of the urinals. They don't show up two to a urinal when there are empty ones. They prefer to have one to themselves. And when even the second spot at each urinal is taken, the knee-clenching next men in line are forced to go upstairs to another bathroom.

But while each orbital has an even number of slots—two—each atom does not have to have an even number of electrons. That means sometimes electrons have a slot in the orbital all to themselves. They are unpaired. The French call them "celibate."

This is where magnetism comes in. When a material is made up of atoms with one or more unpaired spinning electrons, the atom itself creates a tiny magnetic field. But in some unusual substances, a majority of those solo electrons can be made to spin in the same direction, lining up, magnifying the field in a larger material, making a sphere of influence greater than themselves. In most substances, that field is weak and passing and can be measured only by sensitive instruments. In some, the fields within a substance organize themselves in such a way as to cancel each other out instead of amplifying each other. But a few atoms can retain a strong magnetic field. The most common are iron, cobalt, and nickel. The iron atom has four unpaired electrons in its outermost rank of filled orbitals. Cobalt has three and nickel, two. When those elements combine with others to make such materials as magnetite or terracotta or basalt, the magnetic field within the material can last a very long time.

Because unpaired electrons spinning in the same direction create a magnetic field, it makes sense that a magnetic field itself flows in a direction. It does. As with the orbitals, scientists use everyday planetary imagery to describe this phenomenon. They say that a magnet has a north and south pole, where the field is strongest, and that the field travels from north to south.

Magnetic fields moving in the same direction repel each other and those moving in the opposite direction attract. It's the same idea as the positive and negative charges of the protons and electrons within the atom's structure: Opposite charges attract, like charges repel. So when you try to stick a south-facing magnet to a north-facing one, they click together, making the field bigger, magnifying it. But try to put two souths or two norths together and the magnets push each other away. They resolutely refuse to join. This is the fundamental push and pull of the magnet. It is their strong, invisible fields that are doing the pushing and pulling, the same fields that the universe is made of.

Along with direction, a magnet also has a strength or intensity. As you can imagine, if you have a single atom, the field is pretty weak, no matter how many unpaired spinning electrons there are. Gathering a lot of atoms with unpaired spinning electrons together makes a stronger, or more intense, field. So a larger magnet is stronger than a small one. And joining two magnets together, like the south- and north-facing ones we just talked about, creates a more powerful magnet. It makes sense. You've got more of those unpaired spinning electrons all pulling in the same direction.

To fully describe a magnet, or the field it creates, it follows that you need to be able to talk about both direction and intensity. Mathematicians call something made up of two components a vector. We use the language of vectors when we talk about velocity, which is direction as well as speed. So, the car's velocity is 100 kilometers an hour to the northeast. That's different from saying that the car is traveling northeast (just direction), or saying that the car is traveling 100 kilometers an hour (just speed).

The big picture is that if the universe had been created without the fundamental force of electromagnetic interactions, it would be a profoundly, unimaginably different place, right down to the structure of every atom.

CHAPTER 3

parking in the shadow of magnetism's forgotten man

Kornprobst was walking around the perimeter of the property on rue de Rabanesse, savoring Brunhes's role in transforming the science of magnetism. But he wanted to make it clear that he was not an expert on magnetism, just on Brunhes. I was not fooled. There is a tradition among scientists to disavow knowledge of a field unless they have published scores of papers on it. Kornprobst knew much more about magnetism than all but a few hundred people on the planet.

As a young scientist, he fell in love with the Earth's mantle, the thickest of the planet's four layers. It is named after a cloak because it encloses the spherical outer core and inner core. Most of the mantle is enclosed in its turn by the rocky crust. There are a few exceptions where it pokes through. "It was Morocco," Kornprobst said, sighing, a dreamy look in his eye. "There is a splendid piece of mantle there."

The mantle is a highly pressurized mass of silicon-based material— the same element we use to make computer chips—heated by radio-active energy. It moves sedately all the time, more slowly than the core. And the crust, formed of hard tectonic plates, moves yet more slowly still over the top of it, shifting the continents and oceans millimeter by millimeter. Occasionally, the plates rip apart or suck each other under, which can result in earthquakes or volcanic eruptions. Like the inner and outer cores, the mantle's job is to get rid of some of its heat, which is the point of volcanoes and earthquakes. That impulse to shed heat is also the agent behind the Earth's own magnetic field, the one generated within the core. The planet's heat moves from the inside out: from the inner core to the outer core and into the mantle and then the crust. To understand the mantle is, therefore, to have some understanding of the core and of the magnetic force itself.

When Kornprobst was head of the geology department at Université Blaise Pascal in Clermont-Ferrand in the mid-1980s, he established a research center on volcanism and magnetism that put the university at the forefront of the field globally, part of the long love affair French academics have had with the discipline. For ten years, ending in 1998, Kornprobst was the director of l'Observatoire de Physique du Globe in Clermont-Ferrand. It's the same job Brunhes held when he lived in this tower on rue de Rabanesse in the first years of the twentieth century. And this tower was the original Observatoire de Physique du Globe of Clermont-Ferrand.

At the base of the tower was a thick plank door, so low as to encourage stooping. Brunhes must have entered his home and observatory by that door, walked up the six flights of stairs; he must have stood behind the railing of that round turret and thought about both the stars of the universe and the roiling of the planet's core. Some of the trees still standing on the land looked old enough to have lived there when Brunhes and his family did.

A sign hung from the fence. Despite the fact that the structure had been designated a French historic monument in 2009, the property had been sold and was slated for redevelopment into forty-two housing units. Kornprobst set his mouth in a straight line. Again the chin jutted out. He had already helped to organize a student protest about the plan. He wasn't about to stand still for the desecration of Brunhes's first observatory.

The slender Rabanesse tower in Clermont-Ferrand started out as a link in a European chain to gather information about weather and atmospheric conditions, Kornprobst explained. Originally, it had nothing to do with magnetism or with changing the course of science.

It was connected by telegraph line to the world's first mountain meteorological observatory on top of the Puy de Dôme volcano outside Clermont-Ferrand. The volcano is the most famous in the ancient chain of volcanoes scattered through central France, so steep and forbidding that it once formed part of the route for the Tour de France bicycle race. Even today, it is an overwhelming presence in the city, an all-seeing giant. On a clear day, you can see the observatory perched on the volcano's summit, still collecting data about atmospheric conditions.

In turn, Rabanesse was joined to Paris, forming a network of meteorological observations from the mountain to the plains to the capital and from there to the rest of Europe. But Brunhes was not only interested in meteorology. Born in 1867, he came from a remarkable family, the oldest of seven intellectually gifted children, according to a short biography of him published in 1999. Perhaps he wanted to make his own mark.

His father, Julien, was the son of a master shoemaker in Aurillac, about two hours' drive south of Clermont-Ferrand. Julien made the leap from the trades into academia, eventually studying in Paris and becoming a physics professor and then dean of science at the University of Dijon.

Both Bernard Brunhes and his younger brother Jean followed their father into the sciences. In fact, after Julien died in 1895, Bernard took his place on the Dijon faculty. But it is Jean, a geographer, who is the better-known brother, famous for inventing the term and discipline of "human geography"—a social science that examines humans' interaction with the environment.

The impression that emerges of Bernard Brunhes from his published papers and the biography is of a frail but driven man, a fervent Roman Catholic and an idealistic social reformer. He and Jean traveled to Rome to meet Pope Leo XIII at the Vatican in 1892, part of a delegation of young Christian socialists. The brothers were inspired by the 1891 papal encyclical *Rerum Novarum (About New Things)*, the first to explore the conditions of the working classes. It is the foundation for the modern Catholic understanding of social justice and climate change, including the work of Pope Francis. Motivated by the encyclical, the Brunhes brothers gave night courses to blue-collar workers.

Academically, Bernard Brunhes was a polymath, like so many scholars of his era. His interests ranged from optics to acoustics, from electricity to thermodynamics to X-rays, from horticulture and botany to what today would be called environmentalism. But in 1900, after he was named director of the observatory in Clermont-Ferrand, as well as a professor of science, he enthusiastically switched gears: The emerging discipline was geophysics. It's a broad field, taking in the shape and structure of the planet and its atmosphere. But it also encompasses seismology, gravity, volcanism, and magnetism.

Brunhes renovated the observatory on the Puy de Dôme, making it bigger, installing a gas motor to run electricity—a brave innovation—and sending his apprentice Pierre David to live there full-time. As for the Rabanesse tower, Brunhes reckoned that it was in the wrong place to produce the best meteorological data, so he devised a plan to build a magnificent new building several kilometers away.

Brunhes's ardor, as he sifted through weather data, ran the Rabanesse tower, and renovated the mountain observatory, turned to the magnetism of rocks. Basic questions were unclear. How did rocks get a charge and hold on to it? How could you tell when they had become magnetic? What did magnetic rocks tell you about the workings of the planet?

By this point in our day's schedule, Kornprobst was driving his little Renault toward the empty parking lot of Les Landais. This was the grand building Brunhes commissioned early in the twentieth century to be built on what is now the main campus of Université Blaise Pascal. He leapt out of the car, looked around possessively, and slowly filled his lungs with the scent of cherry blossoms. Les Landais is a handsome two-floor red building with banks of windows surrounded by a bewildering assortment of cherry trees and a couple of outbuildings. Opened in 1912, it was a dramatic step up from the tiny Rabanesse tower downtown. Kornprobst had been director at this building for nine years, and it was the scene of many of his own triumphs: a Doppler radar that could determine the speed of particles in volcanic clouds, and a seismological observatory with one of France's dozen seismometers, providing up-to-the-minute measurements freely available to the public and scholars in case anyone wanted to track potential earthquake activity.

The main building had been turned over to the French geological survey—the Bureau de Recherches Géologiques et Minières (BRGM)—which collects information about such things as water levels, boreholes, and seams of coal, lead, and uranium in the region. Its latest director had started there only a handful of weeks earlier and Kornprobst, with his customary efficiency, had called ahead to inform her of his visit. He marched up to the door. "Kornprobst," he declared when she arrived. She looked startled at his vehemence, but showed him around.

He was taken aback, perhaps horrified by all the renovations in his former office. The luxury! A wide desk. Magnificent views. From her office, he pointed to the next stop on our itinerary: a huge new building just across a field, with hardly any parking at all. We would leave the car in the parking lot of Les Landais and walk, he said.

Now the director was soothing. Surely Kornprobst could brave the parking stress? We went back to the car. He drove up to the imposing modern cement complex and, rather defiantly, parked in front of it in an illegal spot. It was the third and current home of l'Observatoire de Physique du Globe de Clermont-Ferrand, Kornprobst's pride and joy.

He can trace the observatory's history from the Renaissance tower to Les Landais to this cement behemoth. Among all the physicists who have ever lived in France, just Kornprobst and Brunhes have held the same two important roles: directors of l'Observatoire as well as the guiding minds behind building more modern incarnations of it.

That affinity has led Kornprobst on this quest to make sure the world—or at least the observatory's students and staff in Clermont-Ferrand—remembers Brunhes. It's a strangely hard slog. The scientific memory is usually precise and generous. Kornprobst gestured to an imposing stand-alone panel directly in front of the building. This was the panel he had written about in *Eos*. It was made of brilliant turquoise-colored enameled lava—the lava part is an in-joke among volcanologists and the paleomagnetism crowd—set on two sturdy legs. It celebrates the centenary of Brunhes's find in sharp white script, featuring a side-view portrait of Brunhes in relief.

Brunhes is depicted as a slender man with a carefully trimmed, pointed beard, long neck, and crisp, high collar reminiscent of the fin-de-siècle. This panel and a plaque inside the building featuring the same portrait are, so far, the only two formal monuments in France to Brunhes. Despite Kornprobst's efforts, Brunhes is still a forgotten man of physics.

Gleefully, Kornprobst motioned to the depiction of a compass in the panel's bottom right-hand corner. Four arrows, one pointing to each of the four ends of the Earth. Except in this version, east and west are just where you'd expect and north and south have changed places. It is a puckish reminder of what Brunhes found but also a little-known modern scientific truth: Today's north is actually in the south. That's because in magnetic nomenclature, the pole from which the magnetic field flows is its north. The receiving pole is south. Today, the field flows from what we call the South Pole.

For the entire time *Homo sapiens* has been trying to unravel the mysteries of magnetism, those magnetic poles have been on the opposite sides of the Earth from what we imagined.

CHAPTER 4

into whose embrace iron leaps

Because magnetism is ultimately about how the planet, its geological features, and even its species came to be, new findings have often kicked up against religious orthodoxies. Frequently, as Brunhes found out, they have challenged scientific ones too. Some of the investigators have put reputation, jobs, freedom, and even their lives in jeopardy when their discoveries called into question the teachings of the day. For generations, it has taken a countercultural imagination to puzzle out the meaning of magnetism.

The very name "magnet" has ancient roots in the arts. It goes back to the classical Greek poet Homer, who wrote his famous epics, *The Iliad* and *The Odyssey*, in about the eighth century BCE. Those epics, in turn, were based on the tales of oral poets whose work came even before the invention of the alphabet. Homer writes about the mythical hero Magnes, a son of the Greek god Zeus, who was king of a region of Thessaly in central Greece. It was named Magnesia after him and his people, the Magnetes. One of the minerals common in the lands of the Magnetes was a substance that came to be called magnetite, after the people, their land, and their hero king.

Magnetite, an iron oxide, is a naturally occurring permanent magnet, which means that enough unpaired spinning electrons in its molecules stay lined up in the same direction to keep its field strong. For centuries, magnetite has also been known as lodestone, a word that has crept into literature as a metaphor for a power that guides one's life, or a moral reference point. Modern analysis shows that Thessaly, in Greece, is home to rare pure compounds of magnetite, which means that they were unusually strong natural magnets.

Another tale, told by the Roman author Pliny the Elder in his first-century encyclopedia *Naturalis Historia*, or *Natural History*, is that a shepherd named Magnes discovered the charged stones when the metal in his shoes and staff stuck to them on a mountain in either Asia Minor or Crete. His name became the name of the substance "into whose embrace iron leaps," Pliny writes.

But why does the iron leap? Early Western philosophers toyed with two main explanations, as the historian A.R.T. Jonkers chronicles. One camp held that magnets were drawn together, just as living creatures were, by an affinity or sympathy for each other. The other believed that the attraction was mechanical, born of actual particles or emanations or "effluvia."

Thales of Miletus, who lived in the sixth century BCE, was in the sympathy camp. Known as the first Western philosopher, he was a gifted mathematician and astronomer who discovered the constellation Ursa Minor and correctly predicted the solar eclipse of May 28, 585 BCE. His ideas survive because Aristotle and others wrote about them. He's credited with helping lay the foundation for the modern understanding of the world, resting on steely-eyed observation rather than dogma.

For all his philosophy, Thales could be a practical man. Jeered at by wealthier neighbors in the Greek colony of Miletus on the coast of modern-day Turkey for living in poverty for the sake of his science, he played a sly joke. Figuring out one winter that the weather was shaping up to produce a good crop of olives the next autumn, he raised enough money to rent every olive press at bargain prices. When the olives ripened, he rented out the presses at a handsome profit.

His innovation, Aristotle tells us, was to say that magnets have souls, which is why they can make iron move. It was a bold rejection of the prevailing idea that only the gods could make things happen and a challenge to the worldview that substances and people were the pawns of the gods. Instead, understanding the world meant looking around you, then making theories and testing them—the basis of scientific experimentation. There is no record of whether Thales suffered punishment because of his ideas. Two clues suggest that he did not: He is reported to have died as an old man after collapsing at a gymnastics meet, rather than being imprisoned or shunned, and his school survived long enough to produce many other innovative philosophers.

Empedocles of Acragas, a flamboyant Sicilian fond of wearing bronze slippers, who lived in the fifth century BCE, was in the opposite camp. He held that iron released physical effluvia, or vapors, from its "pores." These vapors were drawn to the lodestone, dragging the iron helplessly in their wake. This physical explanation for magnetism hopscotched from one philosopher to another across the next several centuries, morphing slightly in each new iteration. In the fourth century BCE, Democritus was thinking about theoretical "atoms" that fitted together with balls and sockets or hooks and eyes. He said that bits of iron were drawn into lodestone, forcing the iron to connect to it. A century later, Epicurus attributed the magnetic force to a mysterious circular connection.

Lucretius, the Roman who wrote De rerum natura (On the Nature of Things), in the first century BCE, was also in the mechanical bloc. His only known work is a 7,400-line poem composed in the dactylic hexameter that both Homer and Virgil popularized. It expanded on Democritus's revolutionary idea of the atom. Lucretius contended that everything in the universe, including humans, is made of small bits—atoms. Not only that, but he said that the universe and its creatures have evolved over time. This went much further than Thales's ideas. It was an outright rejection of the notion that the gods had created the world. Within this astonishing work, Lucretius described the lodestone as "the stone men wonder at," as a Victorian translator put it. He said that air circulating between iron and lodestone created a vacuum, which drew the two together.

Lucretius was lyrical if not comprehensive about the lodestone. And his work lay forgotten in just a few moldy handwritten manuscripts buried in European monasteries until it was rediscovered in the fifteenth century. Once again, it was incendiary. It would eventually influence some of the world's most revolutionary modern scientific minds, including Galileo Galilei, Charles Darwin, and Albert Einstein.

But for some of the early magnetic theorists, magnets did not have stories to tell about the planet's birth or its future or the place of the gods. They were not ideologically controversial. Instead, magnets were curiosities or tools. Hippocrates, who founded Western medical thought in the fifth and fourth centuries BCE, said that bleeding could be staunched by binding a piece of magnetite—either whole or powdered—directly to the body. It was a natural progression from there to outright magic. By the fourth century CE, the author of the *Orphic Lithica*, a poem about the magical use of stones, wrote that magnetite could produce the pull of passion, whether human or divine. In other words, it was the longed-for key to desire.

Its use as a navigational tool started early too. The Chinese appear to have begun developing intricate compasses using magnetite in the centuries before the common era. The Chinese called magnetite the "loving stone," presumably for the same reason the French call magnets *aimants*, or "lovers." In her book *North Pole, South Pole*, Gillian Turner, a physicist and historian of magnetism, describes the replica of an instrument likely used to help the Chinese lay out villages in harmonious directions. It featured a magnetite spoon symbolizing Thales's Ursa Major constellation. The spoon rested on a bronze or wooden plate representing the heavens. Both the constellation and the heavens sat on a square plate, which was the Earth. The bowl of the spoon pointed north. Curiously, the Chinese appeared to use south as the primary reference direction rather than north.

By the early twelfth century, Turner writes, the Chinese had become brilliant at making navigational compasses by stroking iron needles with magnetite, therefore forcing the iron needle's unpaired electrons to line up temporarily in the same direction. They suspended the magnetized needles on silk threads or otherwise allowed them to point north-south, a feat not perfected in the Western world until decades later, when mariners began to rely on magnetized iron needles in their compasses. These seamen were known as "lodesmen."

But it was a medieval French engineer who was the father of modern magnetism, and some say, of modern scientific investigation. Pierre Pèlerin de Maricourt, also known as Petrus (Peter) Peregrinus, was a thirteenth-century scientist—possibly a knight—born in Picardy, the charmed region of northern France famous for its Champagne vines. There is no record of his early life, but he lives on in the annals of science for the 3,500-word letter he wrote to a friend on August 8, 1269, several copies of which survive. It is a treatise on magnetism written in Latin, the language of scholars and the upper class, and it contains results of some of the first experiments recorded in the history of science.

Peregrinus seems to have been a tinkerer. He was certainly a builder of machines who may have studied at what was then the new University of Paris. The sobriquet "Peregrinus," which literally means "a man who wanders" in Latin, signals that he was a crusader, a noble soldier fighting in one of a slew of the medieval era's military campaigns to further the interests of the Catholic Church, sanctioned by whoever was pope at the time.

Peregrinus wrote his letter while he was in the employ of Charles of Anjou, who was waging war against the mainly Muslim inhabitants of the hillside town of Lucera in Italy, a geopolitically important site in the Middle Ages. Charles, the brother of Louis IX of France, was one of the most ambitious of medieval European noblemen— and that was a rather high bar in those warlike times. Peregrinus was using his knowledge of warcraft to help with the prolonged siege, building fortifications around the French camp, laying mines, and overseeing siege engines such as catapults to lob missiles of fire and stone into the fortified city.

Peregrinus obviously had some time on his hands, because there, in the shadow of war, he began thinking about the Greek mathematician Archimedes, who lived in the third century BCE and who was reputed to have made an ingenious, three-dimensional, Earth-centered model of the solar system, possibly in bronze. What if such a sphere could keep moving forever? Peregrinus wondered, imagining what today we would call a perpetual-motion machine, or dynamo. That's when he began to muse about the magnet. You can almost picture him in the soldiers' camp in the blistering heat of Italy's August sun, distracted from war by the puzzles of the universe. And so he devised experiments to test the properties of the magnet.

It's hard, today, to reconcile this scientific urge with the society he lived in. Printed books did not exist. Paper was uncommon. Archives, including records from some of Charles's other battles, were made on illuminated manuscripts created from painstakingly scraped sheepskin vellum. Brilliant reds on the manuscripts came from heating alchemical staples such as mercury or sulfur. Blues were from lapis lazuli, carried to Europe from the Middle East by camel and then ground into fine dust, fixed with egg or gum made from the boiled skin of a mammal. And while Greek and Roman architects and artists may have recognized linear perspective, by Peregrinus's age, images were back to being two-dimensional. The knowledge of how toxic the art supplies could be was centuries in the future.

Only about a dozen universities existed at that time anywhere in the world. The University of Paris, later called the Sorbonne, was barely more than a century old. The works of Aristotle and eventually Plato, long forgotten, were newly on the rise, part of a pulse of interest in education and culture that some scholars trace back to the first crusade in 1095. That crusade, and those that followed, introduced Europeans to the intellectual mysteries of the Greek and Islamic worlds, so when Latin translations of the ancient Greek philosophers began to appear for the first time, the best medieval minds pounced on them, parsing them intimately for insight into how the world worked.

But at that time, science was philosophy, not observation; idea, not experimentation.

The Bible was the most important book, the final word on the disciplines we now call physics, chemistry, geology, and biology. If the Bible said something, it was the word of God and therefore true. The official Vatican interpretation of the Bible was infallible, even if scientific observations seemed to contradict its interpretation. Peregrinus's duty was to use science to bolster ideological belief.

It was clear that Peregrinus knew he was onto something controversial. Most people of the day thought of magnetism as a fleeting magic: Sometimes it was there and sometimes it wasn't. You couldn't count on it. Or they saw it as a troubling, forbidden, quasi-sexual affinity that led one item to be drawn irresistibly to another, something that might not be able to be withstood or controlled, something that metaphorically represented the act of sexual congress itself, possibly the work of the devil. The idea that magnetism could be a fundamental phenomenon of the universe could not have been further from the public mind.

"The disclosing of the hidden properties of this stone is like the art of the sculptor by which he brings figures and seals into existence," Peregrinus wrote to his friend. "Although I may call the matters about which you inquire evident and of inestimable value, they are considered by common folk to be illusions and mere creations of the imagination."

What hidden properties? Peregrinus was the first to figure out that a magnet has two poles. He was one of the few early investigators to note that a magnet repels as well as attracts. But to Peregrinus, the idea of poles did not contain the idea of movement, or of a field. He couldn't have imagined unpaired spinning electrons in an atomic array. His explanation was that the magnet carried within itself a replica of the heavens, by which he meant the stars that point to geographic poles, aligning with the axis of the planet. They're called sailors' stars because sailors have used them to navigate for hundreds of years and occasionally do so today.

Peregrinus gave his friend instructions on how to find a lodestone's poles, the results of his astonishing experiments. During what was presumably a quiet time at the siege of Lucera in the summer of 1269, he writes that he first placed the lodestone in a small round wooden bowl, then placed that bowl in a large vessel so full of water that the small one floated. The north pole of the stone steered the small bowl toward the north pole of the heavens, and the south pole of the stone to the south pole. "Even if the stone be moved a thousand times away from its position, it will return thereto a thousand times, as by natural instinct."

Schoolchildren have now for generations conducted experiments similar to Peregrinus's. They rub a needle vigorously across a magnet, which aligns the unpaired electrons in the needle's iron content temporarily with the magnet's poles, and then they place the needle on a cork or a bar of soap in a bowl of water. The needle's north will then align with what we call magnetic north, and the south with the south.

The details of where and how Peregrinus conducted his experiments have not survived. But there are a few clues to set the scene. During the Second World War, British aerial reconnaissance photographs of that part of Italy revealed buried sites likely from Charles's siege of Lucera and from much earlier Roman and Neolithic settlements. Subsequent archeological excavations unearthed evidence of a substantial medieval military camp outside the walls of the besieged town. Pottery found there from that period includes large glazed dishes with ringed bases painted green, yellow, or brown; often featuring a lively painted figure at the center, usually a mammal, bird, fish, or human; likely large enough in which to float a small wooden bowl containing a lodestone. Illuminated manuscripts describing Charles's military actions at this time depict men with chins meticulously shaved, hair cut in a fringe high on the forehead, falling to below the chin and then cut across the top of the shoulder in an unswerving line. In battle, their heads and necks were protected by a cowl of fine chain mail. Helmets were rounded. The most noble of the men, likely including Peregrinus, wore mid-calf-length garments dyed in brilliant solid colors, girded with a leather belt, sword slung at the hip. Crossbows were in common use. Soldiers who could not afford swords or crossbows showed up with spades and pickaxes. Military engagements in the thirteenth century were primarily affairs of the cavalry, so horses were a feature of the camps and therefore so were farriers and smiths. How did Peregrinus manage to steal away to his scientific endeavors amid the cacophony and odor of a camp such as this?

Shortly before Peregrinus finished his magnetic letter in 1269, Charles of Anjou, tired of waiting for more than a year for the inhabitants of Lucera to succumb, cut off their food supplies, making sure that a thirty-mile radius around the town was devoid of animals and other foodstuffs. Less than a month before Peregrinus wrote his letter, Charles, who had already instituted a thorough conscription, stepped up his reserves, ordering in five hundred lances for men on horseback and another five hundred for those on foot. He hired one hundred carpenters, plus brick makers and wall builders, and laid in hemp, rope, chamois leather, and iron and grindstones to sharpen the weapons. By the time Peregrinus finished his letter on August 8, the inhabitants of Lucera were reduced to eating grass. They surrendered before the end of that month, starved into submission. Three thousand were slain. Charles boasted that they had prostrated themselves on the ground before he had them slaughtered.

Peregrinus mentions none of this drama. Maybe he was hunkered down in the mess tent by wooden bowls and water vessels, but, alas, those details are lost.

We do know Peregrinus didn't stop with the easiest experiments. He discovered that even if you cut the lodestone in half or in progressively smaller pieces, each piece was a new magnet with the same two poles. It would have been more logical to think that if you cut a magnet in half, one pole would remain in the first half and the second pole in the second half. But not so. And it didn't matter how small you cut the magnet; each piece still had two poles.

His experiments also showed that north poles attracted south ones and repulsed other norths. South attracted north and shied away from south. This was revolutionary stuff. Then he used his findings to create an early version of the round compass, a magnetized needle surrounded by a circle divided into 360 degrees. It could be used to describe where one was in the world.

But while all those discoveries were shocking and new, Peregrinus's greatest finding was that the lodestone carried what he called natural instinct. That meant that the lodestone's powers were not ephemeral but constant, and, in Peregrinus's understanding, irrevocably linked to the power of the stars. He was still a long way from being able to peer inside the Earth and understand the forces that create magnetism or—more revolutionary still—posit that those forces could reverse the direction of the magnetic flow, but his was the first effort to establish that magnetic power surrounds us all, invisible and inescapable.

revolutions on paper

When Kornprobst was a graduate student at the Sorbonne in Paris in the late 1950s, his professors used to save their highest ridicule for two theories: that the continents drift and that the magnetic poles reverse. Today, Kornprobst said, craning his neck to point up a hill behind a Citroën car dealership in the town of Boisséjour near Clermont-Ferrand, both are geological gospel.

The hill was the site of another piece of the Brunhes puzzle. Shortly after assuming the directorship of the observatory in Clermont-Ferrand, Brunhes had read three key scientific papers that had launched his quest to understand more about magnetism. Like other scientists of his generation, he knew that the magnetic field was shifty. It was thought to be composed of three parts: declination, inclination, and strength. (Today, they are compressed into direction and strength, but declination is still noted on any high-quality map.) Any point on the planet could be described by those three magnetic coordinates. It was more complex than the typical two-dimensional geographic mapping coordinates of latitude and longitude. The problem was, magnetic coordinates changed slightly over time.

Declination is straightforward. Compasses point home to the magnetic pole along magnetic lines of force that converge at the poles. But beginning in the eleventh century, the Chinese scientist Shen Kuo realized that the spot the compass points to is different from the Earth's geographical pole. The geographic North Pole, for example, can be found directly underneath the North Star. Shen wrote about this in his 1088 chronicle *Meng Chhi Pi Than (Dream Pool Essays)*. By the early fifteenth century, European mariners knew this too. The angle of difference between the geographic pole and the magnetic line of force is known as declination. By convention, if you are east of geographic north, your declination is positive; west is negative. Declination changes depending on where you are on Earth. It also changes if you stay put but measure it over time. Not only do the magnetic poles themselves shift around, but the magnetic field lines that the compass responds to do too. It's counterintuitive, especially if you've done those magnetic experiments with iron filings shaken onto a piece of white paper laid over a bar magnet. The filings arrange themselves with spooky precision along the magnet's lines of force, intersecting at the poles. But the Earth's field lines are not tidy like the ones on the white paper. They are stretchy and prone to wild distortions. Over decades, those changes could be dramatic. For example, modern reconstructions show that declination in London in 1653 was positive. By 1669, it was negative. After gyrating greatly in the ensuing years, by 2018 it looks headed to become positive again.

Inclination, or magnetic dip, was a later discovery. Measurements began sporadically in England in 1576, and the idea gained prominence after the publication in 1581 of *The Newe Attractive*, a pamphlet written by the Elizabethan mariner Robert Norman. After having spent a couple of sharp-eyed decades at sea, Norman became a master instrument maker in London. His revolutionary work on magnetic dip stemmed from years of experimentation with the sea compasses he was making. He discovered that if you have a compass needle moving freely in a sphere and you point it at the horizon, it will be pulled either up or down by the Earth's magnetic force, depending on where you are compared to the magnetic equator. In London, Norman measured the dip at an angle of 71 degrees, 50 minutes. At the equator it doesn't dip at all, remaining horizontal. At what we call today's North Pole, it will point straight down, and at the South Pole straight up.

The fact that things changed so much implied that the Earth's magnetic force was a vast entity moving to its own inscrutable rhythms. Scientists of the nineteenth century were determined to crack the code. One goal was to measure regional variations in magnetic fields over time, to reconstruct the Earth's magnetic life back through the centuries. That's where Brunhes entered the picture.

The first of the three critical papers that caught Brunhes's attention was by Macedonio Melloni, an Italian physicist who founded the Vesuvius Observatory just outside Naples in 1848, the same year revolutions were bubbling over in cities around Europe. That meteorological observatory was set up in part to monitor the activity of Mount Vesuvius, the volcano that had erupted in 79 CE, destroying the Roman settlements of Pompeii and Herculaneum, killing at least 1,000. Today Vesuvius is one of the few active volcanoes in Europe. Melloni selected the observatory's site, designed its building, and chose the instruments. But months after it opened, Melloni was fired and only just avoided being banished from Naples. He was caught up in a mass ouster of academics following an insurrection against King Ferdinand II of Naples. A military letter recently uncovered in the Naples state archives disclosed that political leaders had deemed Melloni "bad" because he was friendly with some of Europe's ultra-liberal thinkers and radicals. Among them were famous scientists, including the great British physicist Michael Faraday.

Melloni was no stranger to bitter scientific controversy. He had been drummed out of Paris's scientific circles earlier in his career for findings that linked heat and light. By 1834, that same work captivated London and Melloni became one of the most celebrated physicists in Europe.

After being fired from the Vesuvius Observatory for being a radical liberal, he took himself to Portici, just outside Naples, and re-ignited an earlier interest in measuring the magnetism of volcanic rocks from Italy and Iceland. Melloni had developed a simple process reminiscent of the early Chinese compasses: a pair of needles, nine centimeters in length, magnetized and hung one above the other by silk thread. When he passed a piece of lava near the upper one, he could measure whether it attracted the needle away from the orientation of the lower needle, and by how much.

He found that all the lavas made the needle move. And he went on to make a bold contention: The lavas had captured the precise magnetic coordinates of the spot on the Earth where they were when they cooled. The idea was that a piece of lava laid down in Italy would have different magnetic coordinates from one in Bolivia. We could say that the electrons had developed a magnetic memory, aligning in a sort of magnetic fingerprint that helped identify the declination and inclination and intensity of where they were located.

Melloni went further. In his lab, he heated lava rocks until they were red hot, at which point they lost their original magnetic memory. When they cooled, they acquired a new one. The flaw in his research was that he didn't systematically determine whether a batch of lava showed the same clear magnetic orientation across its flow. His results were fascinating but not conclusive. He died in a cholera epidemic in 1854.

By 1899, Giuseppe Folgheraiter had taken Melloni's findings to another level. This was the second key paper Brunhes read. Based in Rome, Folgheraiter examined archeologically dated terracotta clay pots from ancient Greek, Roman, and Etruscan civilizations and found that they retained a strong magnetic orientation, even over many centuries. He surmised that they held the coordinates of the magnetic field from the time they were baked.

The third important paper was by Pierre Curie, the French physicist who went on to win the Nobel Prize with his wife, Marie Curie, and Henri Becquerel for their work on radioactive substances. Pierre Curie discovered in 1895 that any solid heated to a high enough temperature loses its magnetic properties. The temperature depends on the material and is in the hundreds of degrees Celsius. In a nutshell, the unpaired electrons become excited and confused by all that heat and refuse to line up in the same direction. In certain and relatively rare materials like terracotta and lava, when the atoms cool down again, their unpaired electrons line up once more in a field, taking on the coordinates of whatever field they are in at the time. The temperature that suspends their magnetism is known as the Curie point, and today is an undisputed rule of physics.

Brunhes, sitting in his Rabanesse tower a few years later, put all these pieces together. He was in the perfect place to do it. There he was, nestled in the remains of a string of ancient volcanoes in central France. Obviously, there had been hot lava. Some places also had natural terracotta laid down in a layered sedimentary bed. Terracotta contains iron-based molecules, and Folgheraiter had shown that it retained a magnetic signature. What Brunhes needed was an undisturbed seam of terracotta that had been heated up when lava poured over it, Kornprobst explained to me. It was hard to find. But there was some in Boisséjour, near Clermont-Ferrand. So he traveled there, likely by mule, where the Gravenoire volcano had erupted 60,000 years earlier, and collected a few samples.

One of the car salesmen came out to see what Kornprobst was doing. Kornprobst explained that he was retracing the steps of a famous French physicist who had hacked a small piece out of this hill a century before. The salesman shrugged and went back inside the dealership. In the end, the tiny cube of terracotta clay Brunhes cut out of the hill in Boisséjour didn't tell him much. But, more determined than ever to find more magnetic evidence, he set about trying to find a much better volcanic site.

CHAPTER 6

the earth's magnetic soul

Like so many who have studied magnetism over the ages, William Gilbert did it in his spare time. He was an accomplished medical doctor, and by the time he published his towering work, *De Magnete (On the Magnet)*, in 1600, he was at the peak of his profession. That year, he was physician to Elizabeth I, then nearing the end of her life.

Magnetic theory had come a long way from the thirteenth century, when Peregrinus had developed his idea that the magnet carried a replica of the heavens—with its north and south poles—within its very body. In Peregrinus's understanding, those heavens were perfect and unchangeable and therefore the magnet was too. In the centuries after him, some mathematicians, troubled by the odd variability of compass readings, had begun to propose that the source of the planet's magnetism was not celestial but terrestrial. It was a quantum leap in imagination. It implied that the Earth was not just a lumpen facsimile of something else, but might have its own unique qualities. Could this mysterious source be a magnetic mountain or a lodestone island? Could it lie in the Arctic?

There were good reasons for the renewed attention to magnetism in Gilbert's day. For Peregrinus, parsing the magnet stemmed from intellectual and perhaps practical curiosity. By the time Gilbert began thinking about the magnet, it had become an urgent problem.

The Age of Sail had begun, and with it international trade, battles at sea, and the colonizing of continents far from Europe. It meant traversing the open ocean. No longer could seafarers navigate by keeping an eye out for the coastline, or by sounding the ocean's bottom. It meant a far greater reliance on the magnetic compass. However, as sailors were discovering, the compass was fickle at sea. Not only did declination change from spot to spot, it also changed from year to year. Those trying to explain the discrepancies pointed fingers at shoddy instruments, slapdash steersmen, the heave of the waves during readings, the scent of garlic on a seaman's breath when he passed the compass (a flogging offense because it was thought to interfere with a magnet), and even changes to the magnets within compasses themselves. But whatever the reasons, over the course of a lodesman's career, charts could change enough to make a big difference. Lives, fortunes, and reputations depended on knowing where a ship was and where it would be. The primary way of doing that was describing one's position by the geographic coordinates of latitude and longitude. But that was proving impossible.

The great conundrum was longitude. By contrast, latitude was easy. It was just a conceptual set of parallel lines running east and west around the body of the planet, never intersecting. The only one that cuts the globe in half is the equator, and it can be used therefore as the natural reference point. You could find your latitude pretty accurately at sea by using the sun and stars to guide you with instruments such as an astrolabe, quadrant, or cross-staff. And if you were careful, you could sail straight across the ocean along a single latitude, barring island barriers.

But all longitudinal lines run in great north-south circles around the globe, intersecting at each pole, each one dividing the globe into equal halves. Which was the prime meridian, the reference point? Lines of latitude are the same physical distance apart except at the poles, where the slightly flattened shape of the Earth makes them a little farther apart. So, in the main, each minute of latitude is 1 nautical mile. Sixty minutes, which is 1 degree, is 60 nautical miles or 68 statute miles or 110 kilometers. But lines of longitude are different distances apart, depending on where you are on the planet. On the equator, 1 minute of longitude is the same as 1 minute of latitude: 1 nautical mile. At the poles, where the lines of longitude converge, it is 0.

Because the Earth spins on an axis at the same rate every day, changes in longitude as you sail represent changes in both distance and time. The Earth is a sphere rotating 360 degrees per day. That means every hour it turns 15 degrees. Assuming you know the longitude of your home port and what time it is there, as well as what time it is where your ship currently is, you can tell how many degrees of longitude you have traveled each day. If you know your latitude too, you can also tell how many kilometers or miles you've gone. In those days, sailors had to have superb maths and geometry skills. They also had to be able to read the stars.

The problem was that clocks of that era kept time by pendulum, and they could not do it accurately on a moving ship. It was an intractable problem for four centuries, bedeviling the finest minds in Europe. Therefore, navigators and scientists were trying to use the heavens to navigate more reliably. That meant reading the difference between the stars that pointed to the geographic poles and the magnetic pole the compass pointed to. To them, that meant finding the longitudinal prime meridian. They believed that someone traveling around the world at the same latitude taking declination measurements would find two spots on precisely opposite sides of the globe with a declination angle of 0. Halfway between the two points, always at the same latitude, the declination should be 90 degrees. In other words, if you knew the prime longitudinal meridian and could read declination properly, you should be able to figure out longitude too. At least, in theory. The principle assumes a regularity in the Earth's magnetic field lines that was wildly wrong, as it turned out. The Earth's field lines stretch and contort from pole to pole like elastic bands, not at all in straight lines.

Failing a reference point, they thought, if you had a complete survey showing declination all over the world, you should still be able to find longitude. All it took was a great deal of determined measuring all over the globe, plus maths, laid over some pretty good maps. The race for a longitudinal formula was on.

At the time Gilbert began his magnetic research in the 1580s— just as William Shakespeare was launching his career as a playwright in London—measurements had been pouring in for years. The problem was, the more data there were, the more confusing the case became. It was a beautiful challenge for Gilbert. He was in full, choleric revolt against the teachings of Aristotle, the ancient Greek philosopher whose ideas then held powerful sway at the universities, including Cambridge University, where Gilbert had studied. Aristotle, who died in 322 BCE, taught that the Earth was the dull and unchanging center of a glorious heaven. Four uniform fundamental elements made up the Aristotelian planet Earth: air, fire, water, and earth. By contrast, the heavenly bodies, which revolved around the sun, were made of a superior substance, the fifth essence, or "quintessence." Those bodies had souls or supernatural intellects, unlike the dreary Earth.

Gilbert had been reading Peregrinus and knew of his experiments with magnets. He embraced the bold idea that if you ran yet more complex experiments, you could make more observations about what was going on in the world. This was in direct opposition to the ruling doctrine that if you knew your Aristotle, you knew everything there was to know about the world. You didn't need to look at the world around you, you just needed to read about it in ancient texts. Empiricism, at that time, was next to heresy.

And then there was the folklore. Assertions about the mysterious powers of magnets had expanded over the centuries. Gilbert decried the idea that a lodestone placed under the head of a sleeping woman would force her from her bed if she had committed adultery. Nor could a magnet make a woman like her husband better or induce one to become melancholic or smooth-tongued. It could not cure stab wounds. The blood of a male deer did not restore the strength of a weak magnet. Nightfall did not cancel a magnet's powers. The only magnetic truths could be found in experimentation, Gilbert said.

He rolled up his sleeves and took to his laboratory. His great innovation was to develop little models of the Earth—he called them "terrellae"—that could reproduce the physical phenomena that he observed on the Earth itself. The models were magnetized spheres, likely made of magnetite, and had poles, equators, and even mountainous excrescences. The Oxford University historian of science Allan Chapman notes that today the idea of making a model and experimenting on it is the essence of much scientific practice. Back then, it was outrageous.

That was bad enough. But to make a model of the Earth and then to discover, as Gilbert did, that magnetism worked differently on different parts of it was to contradict Aristotle's contention that the Earth was uniform and unchanging. Gilbert went much further. His careful experimentation convinced him that the Earth itself—"our common mother," as he called it—is a great magnet. The origin of its magnetic force lay in the core of the planet, not in the heavens or on the Earth's surface. Instead, he said the Earth-magnet created an invisible, permanent force that skittered across the planet, interrupted in places by irregularities such as the iron-rich continents. Not only did the Earth have a soul, like the heavenly bodies, but it had a magnetic soul at that,

Gilbert averred. Not inert, it could attract; it could repel. It had power and perhaps even its own inexorable, unimagined strategy.

This was utterly novel. The source of a magnet's vigor had moved from the heavens to the Earth to within the Earth itself. And while Peregrinus had written three centuries before about a magnet's "natural instinct," implying that it was a constant property, Gilbert was saying that the whole Earth itself carried this fundamental power, and that the power was inextricably bound to the Earth's core. The implications were staggering. Gilbert was conscious of being a pioneer, a detective hot on the trail of the planet's unexplored internal secrets "which, either through the ignorance of the ancients or the neglect of moderns, have remained unrecognized and overlooked." He wrote that for the first time in scientific history he was penetrating the "innermost parts of the Earth."

Gilbert's sense of outrage at Aristotle's casting of the Earth as inferior is palpable in his writing:

> It is surely wonderful, why the globe of the earth alone with its emanations is condemned by him and his followers and cast into exile (as senseless and lifeless), and driven out of all the perfection of the excellent universe. It is treated as a small corpuscle in comparison with the whole, and in the numerous concourse of many thousands it is obscure, disregarded, and unhonoured. . . . Let this therefore be looked upon as a monstrosity in the Aristotelian universe, in which everything is perfect, vigorous, animated; whilst the earth alone, an unhappy portion, is paltry, imperfect, dead, inanimate, and decadent.

Gilbert shows all the signs of one in love with his subject. He seems to have viewed magnetism as a noble property, almost as if it gave purpose to the Earth, as he thought the soul did to the body. One chapter of his six-book tome, written in rather tortured Latin, is devoted to comparing electricity to magnetism. It smacks of the need to deal with electricity in order to dismiss it. Gilbert looked at amber and jet stones and the fact that if you rubbed them, they would attract straw or chaff. Today, we would call that type of energy static electricity. In fact, Gilbert gave electricity its name; the word is derived from the Greek *electrum*, which means "amber." But Gilbert takes pains to explain that amber's pull is transitory, unlike that of the steadfast magnet. "Magneticks" and "electricks" were not at all the same thing, he declared. "All magneticks run together with mutual forces; electricks only allure," he scoffed. While Gilbert's conclusions would be found wanting a few centuries later, it's still instructive that this influential champion of experimental discovery would consider magnetism and electricity in the same breath, even if only to dismiss any similarity.

Gilbert was jubilantly aware of how much at odds his new theory was with that of the ancient Greeks. He called it "our doctrine magnetical" and crowed that it was "at variance with most of their principles and dogmas." Gilbert's main aim in spending his time and fortune on magnetic experiments was to topple Aristotelian natural philosophy, replacing it with one he himself had devised. Gilbert's magnetic doctrine lay at the heart of the even broader philosophy he was working on when he died of the plague in 1603. That work, *De Mundo (About the World)*, was not published until 1651 and has not been translated into English. *De Magnete* reads like a technical appendix to *De Mundo*, according to one scholar who has plowed through the later work.

Implicit in Gilbert's findings was an endorsement of the heliocentric work of the Polish astronomer Nicolaus Copernicus from several decades earlier. Copernicus wrote that the Earth spins on its axis and that the Earth also spins around a fixed sun. Gilbert concluded (wrongly, as it turns out) that the Earth spins because of its magnetic force. While Gilbert didn't expressly endorse the heliocentric philosophy, it was embedded in his work. He was a closet Copernican.

These were dangerous ideas. They ran against a seminal teaching of the Bible and, therefore, against the word of God himself. It is difficult today to comprehend how uncritically citizens, not to mention church fathers, of that time viewed the tales in the Bible. They took each story as infallible information. Several decades before Gilbert published *De Magnete*, for example, the Flemish anatomist Andreas Vesalius had published his groundbreaking work *De Humani Corporis Fabrica (On the Fabric of the Human Body)*. Based on Vesalius's many dissections of human corpses, some directly from the gallows, he declared, among many other findings, that the male and female bodies had the same number of ribs. It was a scandal. It ran against the story of creation in the Book of Genesis, in which God took one of Adam's ribs to form Eve. To prove his point, Vesalius mounted resoundingly controversial public lectures throughout Italy to show the crowds what he had found.

Heliocentrism was just as monstrous. According to contemporary interpretations of the Bible, the Earth was the core of God's creation. That meant everything else had to revolve around it. To claim anything else was to claim that the Bible was wrong. Heresy. A few years before *De Magnete* was published, the Italian Dominican friar and philosopher Giordano Bruno tried to enhance Copernicus's ideas by arguing that each star was a sun with its own roster of planets surrounding it. He was brought before the Inquisition, a court set up to defend the Roman Catholic Church from heretical thought. The Inquisition condemned him and, the same year Gilbert's book came out, Bruno was burned at the stake in Rome.

It didn't matter to the keepers of Christian doctrine that new observations of the stars and planets were showing a different truth from the one they thought the Bible told. Theology and science were indistinguishable from each other. Within a decade of the publication of Gilbert's *De Magnete*, the Florentine astronomer Galileo Galilei was peering into the heavens with his homemade telescope, finding four new moons circling around Jupiter and exploring our galaxy, the

Milky Way. In the years following, his telescopic investigations of the skies convinced him that the Earth and other planets revolve around the sun—just as Copernicus had argued in 1543—and Galileo published this idea in 1632 in the form of a dialogue.

The Inquisition took note, as it had with Bruno. Galileo found himself summoned to Rome to face questioning. The fact that Galileo had read and approved of Gilbert's *De Magnete* became one of the four proofs of guilt presented against him. (Copies of Gilbert's book from that era have been found with the Copernican bits chopped out, likely as directed by Inquisitional censors.) Galileo was condemned in 1633, forced to confess his sad errors and then endure the banning of his book. He returned to Florence to live out his remaining few years under strict house arrest.

Gilbert didn't face those same risks. The fact that he didn't overtly embrace Copernican ideas was likely because he was a conservative society doctor rather than evidence of his fears of official prosecution. This was, after all, England, where the Reformist Queen Elizabeth I was head of a Protestant Church. She was not as apt to be concerned about revolutionary ideas as the twitchy Catholics, who were in the process of trying to staunch the flow of defectors to the apostate faith. But at that time, holding an unfashionable view, while not necessarily criminal, could torpedo a lucrative medical career. Gilbert's colleague William Harvey unhappily discovered as much in 1628, when he published his revolutionary finding that the heart pumps blood around the body. Harvey's well-heeled clientele evaporated, despite his status as royal physician.

Gilbert escaped that fate. But he didn't have an easy time of it. He had predicted the fallout in his typically pointed tone:

> Why should I, I say, add aught further to this so-perturbed republick of letters, and expose this noble philosophy, which seems new and incredible by reason of so many things hitherto unrevealed, to be damned and torn to

pieces by the maledictions of those who are either already sworn to the opinions of other men, or are foolish corruptors of good arts, learned idiots, grammatists, sophists, wranglers, and perverse little folk?

The Jesuits led the attack. Gilbert's offense was imagining the primacy of the sun over the Earth. That, they vigorously disputed. The contention that the Earth is a giant magnet was not nearly as controversial. In fact, the Jesuits embraced the idea, even pressing it into service to refute heliocentrism. They said the Earth's own magnetism held it at the heart of the Maker's creation.

For a time, Gilbert's work figured in the burning longitude question. He had decided that declination, or the angle of difference between the geographic North Pole and the magnetic pole, could never change because the continents, which pulled the compass away from the pole, were immutable. Therefore, if you measured declination once, it would always be the same, no matter how often you went back to the same place to measure it. His finding helped propel the drive to measure declination all over the globe, once and for all.

But it was only a matter of decades before that part of his magnetic theory was proven incorrect. In fact, the magnetic force that Gilbert had correctly deduced lay inside the Earth's core was not immutable. Rather, it was, and is, protean.

CHAPTER 7

voyage into the underworld

The electric train to the top of the Puy de Dôme, the sleeping giant of a volcano that overlooks Clermont-Ferrand, was packed with high school students from Brittany. Kornprobst looked around at them, the flower of French youth on a field trip to the mountain observatory that both he and Brunhes had once ruled over. He would be amazed if any of them had a clue who Brunhes was, he confided to me.

We were nearing the end of our first long day retracing Brunhes's historic steps. We had already made a dispiriting detour to Laschamp, a dot on the map a few kilometers from Clermont-Ferrand. This was where, after sampling more than fifty volcanoes in the area in the 1960s, the PhD student Norbert Bonhommet discovered the Laschamp excursion, or near-reversal, from about 40,000 years ago, the last time the Earth's magnetic field was in disarray. For a while, it was thought to be evidence of the last full reversal. Kornprobst's colleague Jean-Pierre Valet has linked the timing of the excursion to the demise of the last Neanderthal. Kornprobst and I had trudged through the snow with a compass, trying to find the spot where

Bonhommet had taken his sample. No luck. We finally gave up and went for lunch: a rich local specialty called truffade, made of sliced potatoes and lots of butter and even more melted cheese, served with red wine.

Now Kornprobst was keen to play host at the scientific laboratory up on top of the Dôme. Brunhes had spent vast amounts of time there, arduously riding up the switchbacking old Roman road to its summit by mule, fascinated with the ruins of a monumental, multi-tiered temple to the travelers' god Mercury built here in the second century CE. It was one of the largest religious sanctuaries of the Roman Empire, visible from the road the Romans built far below.

The temple had been forgotten for centuries, rediscovered in 1872 when workers began excavating the volcano top for construction materials to build the observatory. To Brunhes, who arrived nearly thirty years after that, the discovery represented a potential wealth of information on the magnetism of the large rectangular slabs of pale Dôme lava used to make the temple. His assistant, Pierre David, tested four, finding mainly that they had held their magnetic memory—an important confirmation of their laboratory technique. The two colleagues vowed to do more tests.

But whenever you go up to a volcano's peak you also go down into its terrifying inner reaches, whether you are there for science or worship or art. Nearly half a million visitors a year flock to the top of the Dôme, at least partly because standing on the lip of a volcano, however long dormant, is like standing at the maw of hell. You're not sure what terrors could be right below, waiting to pounce. It is catching a glimpse of the unknowable forces deep within the Earth that hold the power to erase the world.

The Romans likely knew that it was an old volcano. That's probably why they built their sprawling temple to Mercury there, so far above the road. Even today, with the train, you have to want to get there in order to make the trek. To hike back down would take a vigorous hour and a half. But the knowledge that this was once a volcano got lost after the Roman era, along with the memory of the temple. For more than a thousand years, this volcano was thought to be a worn-down mountain like the others in this area. The menace was a faint memory.

Then, in 1751, the French naturalist Jean-Étienne Guettard climbed to the top of the Dôme and saw immediately that it was one of about ninety volcanoes that run across 30 kilometers (nearly 20 miles) of the Auvergne region of central France. He warned that they might wake from slumber and explode again. Modern volcanologists concur. It caused a sensation to find a string of old volcanoes hidden in plain sight in a part of Europe that had been settled for thousands of years. The subsequent intense interest in the volcanoes, known as the Chaîne des Puys, marked the birth of the formal study of volcanoes. Volcanologists made pilgrimages from all over Europe, trying to understand what made the volcanoes spew their lava and when it had last happened, trying to peer down into the bowels of the Earth and back into time.

Like so many elements of geophysics, volcanology represented a challenge to theology. The age of the Earth was of critical, abiding interest to theologians, partly because they believed that if they knew when the world began, they would also know when it would end. Apart from that, to their way of thinking, God had created the world, and his chronicle of that creation was contained within the Bible itself.

This philosophy reached its logical conclusion in the seventeenth century, when James Ussher, the archbishop of Armagh in Ireland, painstakingly went through the Old Testament books to reckon precisely how long before the birth of Christ major events had taken place. He famously put the birth of the planet at early Saturday evening, October 22, 4004 BCE, needing two thousand pages of Latin to explain how he came to that date. Other scholars calculated it to roughly the same period, meaning that they thought the Earth was not quite 6,000 years old. (Modern scientific calculations put it at about 4.6 billion years old.) By the beginning of the eighteenth century, annotated versions of the King James Bible included Ussher's dates in the margins beside the relevant passages, supporting the idea that the Bible was a reliable chronology. The supposed birth date of the Earth wasn't academically debunked until the late nineteenth century, and was still debated into the late twentieth century, long after the field of volcanology had begun, after paleontologists had begun digging up ancient human fossils, and after Charles Darwin published *On the Origin of Species*, outlining the theory that creatures had evolved over time from common ancestors.

On top of that, naturalists of the eighteenth century had dueling theories about what made the material the volcanoes spat out. They were divided between water and fire. The Neptunists, named after the Roman god of the sea, were convinced that basalt, or swiftly cooled lava, was formed at the bottom of the sea as sediment. Volcanoes were formed by underground explosions, perhaps of tar or brimstone. Their rivals, named Plutonists after the Roman god of the fiery underworld, thought lava was molten rock that built up underneath the Earth's surface and finally let loose. Both Neptunists and Plutonists visited the Auvergne to see the Dôme and the other old volcanoes, each faction trying to use evidence found there to prove its theory. The Plutonists eventually won the day.

But there was still the thorny question of when the eruptions of the Dôme had happened. Modern analysis says the lava began to form about 100,000 years ago 30 kilometers (18.6 miles) below the surface at the top of Kornprobst's beloved mantle, partly the residue of heat in the core. It was primordial molten basalt, so superheated that it needed an escape valve, migrating inexorably up to a magma chamber underneath the Auvergne until, about 10,800 years ago, the pressure became too great. The magma chamber burst open, loosing the seething lava, which surged to the top of the volcano and exploded into the air. The blast was so powerful that it destroyed the vast east flank of the Dôme, spilling lava into the countryside. And then, 1,600 years later, after the Dôme had partly grown back, came another massive outpouring from a magma chamber fed by primitive mantle basalt. Some scientists believe there could be prehistoric communities buried in the countryside beneath the ancient lava of the Auvergne, just as there were below Mount Vesuvius.

Today, this dome within a dome is grown over, capped by the Temple of Mercury and, at the very tip, the laboratory, itself topped with instruments resembling giant white fluorescent tubes that can be seen for miles. Kornprobst and I were making our way slowly past the temple, onward to the lab, over the treacherous ice. It was late March, a week before Easter, but the cold pierced to the bone. Kornprobst's cheeks were ruddy, hair windswept. Small piles of snow lay on the ground surrounding the temple's outer walls. Its pale gray slabs—Brunhes's assistant, David, must have tested their neighbors—stood out against the whitened sky. We were in an icy cloud, suspended over portending fire.

I could see why a string of religions across time and space have paid homage to the volcano and the underworld it connects to. Vulcan, the Roman god of fire, gave his name to the volcano and its study. He was the disabled god of the forge, son of Jupiter and Juno, who symbolized both life and death, making thunderbolts for his father. Japan's Mount Fuji, which last erupted in 1708, is a spiritual lodestone, the sacred symbol of Japan itself. Pilgrims climb it at night in order to bear witness from its summit to the sun rising.

That awe has spilled over into the long literary tradition of imagining the inner Earth, a forbidden land of sin and suffering poised for the unleashing. From Dante Alighieri's fourteenth-century *Inferno* of punishment to John Milton's seventeenth-century *Paradise Lost* to Jules Verne's nineteenth-century adventure tale *Journey to the Centre of the Earth*. Verne's heroes descend to the hellish core of the planet through an exhausted two-domed Icelandic volcano, dodging death at every turn, only to ride to the surface again on the boiling plume of a Mediterranean volcano. Or there's a more recent iteration, the American cult television show *Buffy the Vampire Slayer,* whose star spent seven fraught seasons trying to keep the Hellmouth, aka Sunnydale, California, closed so the evils wouldn't come out and destroy civilization.

Kornprobst and I made it to the final stretch of the path around the temple, and then up the slippery outdoor stairs to a single locked door. He rapped on it and a head eventually poked out. Kornprobst introduced himself. He had once been in charge of the whole works and was here to look around.

This scientist was one of the eight or ten who were working in the lab—some temporarily living there—taking samples of the air. It was a far more modern, far-ranging, and technologically intensive process than when Brunhes had run things with his single gas motor. Today, the mountain observatory is part of the World Meteorological Organization's Global Atmosphere Watch, a network of observatories to monitor humanity's effect on the global atmosphere. Jean Brunhes, Bernard's brother, the inventor of human geography, would have approved of the research.

Because the Dôme is the highest of the Auvergne volcanoes and because there is so little heavy industry between the Atlantic Ocean and it, the air quality is unusually pure, Kornprobst told me. The young scientist explained in fluent detail how one of the machines worked, and showed us a raft of that day's findings. One number stood out: The greenhouse gas carbon dioxide was present at 403.197 parts per million—a shockingly high concentration compared to the preindustrial figure of 280. A marker of climate change, this was one of the figures Bernard Brunhes could never have imagined would be so important just a century later.

Finally, Karine Sellegri, director of research of the National Center for Scientific Research's Laboratory of Physical Meteorology and an expert in cloud chemistry, among other things, came to meet us. She was the team leader.

"Kornprobst!" my companion declared, bowing slightly. He explained that we were trying to help the world understand the significance of Brunhes's bold finding that the magnetic field had reversed. Sellegri was apologetic. One small conference room was named after him at the lab, but nobody was told why. His name was not mentioned in courses at the university in Clermont-Ferrand. However, there had been a small lecture on reversals, she recalled, brightening up a little, and she, of course, knew of Brunhes's experiments.

"They were not experiments!" Kornprobst barked. "They were observations!"

Later, over espresso and a biscuit as we were waiting for the train to take us back down the volcano, Kornprobst was reflective. In retirement, he was scientific adviser to Vulcania, known as the European park of volcanism—an amusement park set up to help the public learn about volcanoes and other inner-Earth mysteries. The park was another sign of just how forcefully volcanoes had shaped the psyche of the region, partly spurred by Kornprobst's own fiery determination as head of the observatory to make Clermont-Ferrand an international center of excellence in the discipline. Just a few kilometers from the Dôme, Vulcania's tourist season was launching that evening to great fanfare, and Kornprobst was to be a guest of honor.

As he sipped his espresso, he lovingly tracked the science of the discovery of magnetism back through time, fluidly quoting not only the main figures—Peregrinus and Gilbert, among others—but also landmark papers that had advanced the field. Brunhes was a fulcrum. There was the science of magnetism before him and there was the science of magnetism after, and they were not the same. The next day, we would take the most important journey of all: a trip to try to find the very same seam of rock Brunhes took his fabled samples from, the ones that changed the history of science.

CHAPTER 8

the greatest scientific undertaking the world had ever seen

On June 12, 1634, just as the sun was entering the critical phase of its summer solstice, Henry Gellibrand transported two foot-long magnetized needles and two quadrants to the garden of Mr. John Welles in Deptford, just west of London. He used the instruments to take five sets of measurements in the morning—the sun marking true north, the needle pointing to magnetic north—as the sun marched up in the sky. He took six sets of measurements in the afternoon as the sun marched back down the sky. Then, Gellibrand, a young mathematician who held the coveted position of professor of astronomy at Gresham College in London, did a few sums.

The results transformed the understanding of magnetism. Even Gellibrand, who has been described as a "plodding, industrious mathematician without a spark of genius," could see that right away. The angle of declination in (or at least, near) John Welles's garden in Deptford had shifted by more than 7 degrees in just fifty-four years. That meant declination changed over time even if you stood on the same spot of the Earth and used the same instruments.

It ran against everything the navigators and academics of the day thought they knew. It was one thing to understand that declination changed at sea, where there were so many variables, including the motion of the waves, cloud cover that masked the celestial guideposts, and dodgy geographical coordinates. It was a different kettle of fish to know that right there, in a garden near London, declination was changing. And not only changing but changing fast. Subsequent measurements have shown that the angle of declination in London, then one of the best-measured cities in the world, was on a fast canter, shifting from 11 degrees east in the late sixteenth century all the way to 24 degrees west by 1820.

The incredulity the discovery spawned is almost impossible to fathom, catapulting magnetism into one of the greatest scientific puzzles of the day, right after gravity. In our era, it would be like waking up in the morning to find that time had begun to run backward: A staple of the universe that you had taken for granted no longer worked the same way.

For one thing, Gellibrand's finding meant that Gilbert's idea that the Earth's magnetic field was like a simple permanent magnet could not be correct. The field was in motion. For another, it meant that every single measurement of declination studiously recorded over the years was useless unless the date had also been recorded. It meant that a magnetic coordinate contained within it not only declination and inclination but also the dimension of time. It meant that you couldn't take a measurement and be satisfied that it would be the same years later.

There had been hints before Gellibrand. He was building on the work of his predecessor as professor of astronomy at Gresham College, Edmund Gunter, the brilliant mathematician who invented the slide rule. Gunter had taken what must have seemed to him to be routine measurements of declination, likely in Deptford, in 1622, aiming to verify those taken forty-two years earlier in the same place by William Borough, a seaman and enthusiastic butcher of pirates who became comptroller of the Queen's Navy. Gunter was astonished to realize that there was a discrepancy of more than 5 degrees in the sets of measurements. He put aside the unthinkable: that the magnetic "soul" of the Earth was not fixed. But he carefully recorded his measurements. Gellibrand took up the post when Gunter died in 1626. Eventually, Gellibrand took note of the discrepancy and carted Gunter's own needle to John Welles's garden to repeat the measurements in 1634. He published his findings the following year, naming the phenomenon "secular variation," after the Latin *saeculum*, or age. If we were to name it today, we would call it long-term variation.

In that single academic publication, the discipline of magnetism had undergone a revolution. From the localized, inconstant property of lodestone that the ancient Greeks knew to the perpetual "natural instinct" of Peregrinus's magnet to the permanent global phenomenon Gilbert described, magnetism was now akin to an invisible, capricious, living force within the planet. It meant magnetism must always be on the move, whether you measured it from second to second or across millions of years. What secret power could be buried inside the Earth to make this happen? Where would it lead?

The logical implication to the scientific and seafaring minds of the day was that if magnetic readings changed, there must be a way to calculate how they would change. In other words, there must be a rationale for the changes, and therefore a mathematical formula that would at last allow navigators to use declination to read longitude. This hypothesis meant taking more measurements and trying to figure out what the formula was. The impulse to solve this problem in an orderly fashion was overwhelming.

This imperative set the stage for the fabled exploits of Edmond Halley and for the direct intervention of the British monarchy in the growing urgency to solve the problems of navigation. A physicist who became Britain's second Astronomer Royal, Halley is most famous for the comet that bears his name. He correctly calculated the comet's path through the heavens and predicted that it would reappear in 1758. But being fascinated with the celestial realm at that time meant also being fascinated with navigation and with magnetism. So, in the final years of the seventeenth century, Halley, the son of a wealthy soap maker, took to the seas on the first ocean-bound expeditions undertaken solely for the sake of science, at the pleasure of the Crown.

Despite a dreadful, truncated first voyage on his ship the *Paramore*, a flat-bottomed little pink commissioned specially for him by royal decree, Halley came triumphantly back to London after the second voyage with a list of declination measurements—he didn't measure inclination—stretching across the Atlantic Ocean and down past the tips of both Africa and South America. He had ventured farther into Antarctic waters than anyone before him, calling it the "Icey Sea."

All London awaited his findings. But how to express what he had found in any way that could make sense to the men who steered ships?

Halley had a brain wave. Instead of a table of numbers, he would make a map. He plotted his measurements of declination on the map, and where the numbers showed the same angle he drew curved lines to join them, just as he had plotted the comet's elliptical orbit. He ended up with a document that once more radically changed the way people thought about magnetism. And not only how they thought about it, but also how they saw it.

"A New and Correct CHART Shewing the Variations of the COMPASS in the WESTERN & SOUTHERN OCEANS as Observed in ye Year 1700 by his Ma.ties Command by Edm. Halley," shows swooping new lines laid overtop the familiar grid of longitude and latitude. Some lines indicated that the compass was to be adjusted eastward and others westward, depending on where you were at sea. It was the first published visual image of the Earth's magnetic lines, evidence of what we now know to be a force field of magnetic energy pulsing from the core, spouting from one pole to the other in elastic-band lines, and enveloping the planet. Somehow, the magnetic force now had to be thought of as a volatile, evolving, and constantly moving power that touched everything on the planet.

Seamen needed a little explanation. Up in the top left corner over-top "Canada New-France" and directly underneath "Hudson Bay," Halley wrote: "The Curve Lines which are drawn over the Seas in the Chart, do shew at one View all the places where the Variation of the Compass is the same; The Numbers to them, shew how many degrees the Needle declines either Eastwards or Westwards from the true North; and the Double Line passing near Bermudas and the Cape de Virde Isles is that where the Needle stands true, without Variation." Halley had apparently found Ground Zero of declination. It didn't run along any known longitudinal line, as theorists had lovingly imagined, but instead bisected the Atlantic Ocean between the western bulge of Africa and the eastern flank of South America before veering off underneath Bermuda and right to the shore above Florida.

Today, those same types of cartographical markings are known as contour lines on terrestrial maps, where they indicate different topographical heights. For example, at the base of a mountain, a loop might indicate 100 meters above sea level. Farther up, another loop might indicate 200 meters, and so on to the top. On meteorological maps, these curved lines are called isobars or isotherms, depending on whether they show barometric pressure or temperature. Then, they were called Halleyan lines, and they were an instant hit. Halley later expanded them to parts of the Pacific Ocean with other mariners' data and updated things when he got new information. They were published in some form until the nineteenth century.

Alas, despite Halley's claim to the contrary, his chart was incorrect as soon as he published it. The Earth's magnetic field had already shifted. And the chart was all but useless for finding longitude, except when sailing where Halley's curved lines ran parallel to a coast, and only until the field moved on once more.

As he explored the changeability of the magnetic field, Halley showed that some of its components appeared to be moving to the west, a phenomenon known as "westward drift." To explain it, he developed an intriguing model of the Earth's interior, suggesting a liquid core surrounded by empty space—which he thought might have been home to unknown creatures—contained within an outer shell. He proposed that the inner core had its own pair of poles, in addition to the shell's two, for a planetary total of four. The inner pair rotated more slowly than the outer pair. In simplistic terms, the pairs of poles were fighting each other for dominance, pulling this way and that on the magnetic lines.

Although wrong in several fundamental ways, this model represented an attempt to uncover a comprehensive explanation for the variation in the Earth's magnetic impulse, a step that took Gellibrand's finding that magnetic readings changed over time dramatically further: This was the "why." Halley's new planetary model was also a prophetic attempt to place the cause for the variation within the Earth's liquid core. It was not universally accepted. But suddenly, the quest was on for a hypothesis to describe the magnetism of the whole planet, and what caused it in the first place.

Halley, who died in 1742, made an uncannily accurate prediction: "It will require some Hundreds of years to establish a complete doctrine of the Magnetical system," he wrote.

When he died, a big piece of the geomagnetical puzzle remained unsolved. There was declination, which varied tremendously over time, particularly over the Atlantic Ocean and less over the Pacific. There was inclination or dip, the pulling down or pushing up of the magnetic force on a magnetized needle, compared to the horizon. It changed over time too, but seemingly not as much. Both measurements gave information about the direction of the magnetic force. But what about its strength? In the language of physicists, it was like having knowledge of one component of the vector but not both. It was akin to knowing that a vehicle was heading northwest but not knowing whether it was going 10 kilometers an hour or 100 or 1,000.

Some explorers of exotic latitudes had already used their dip instruments to figure out that the push and pull of the compass needle was stronger the closer you got to the poles. You could tell by applying a mathematical formula to the period of the oscillation of the dip needle as it was pushed or pulled and then returned to where it had begun. But that just measured intensity compared to the strength of the magnet you were using, not actual intensity. The German naturalist and geologist Alexander von Humboldt decided to establish a standard for comparison anyway. Just as the nineteenth century was dawning, he went on an extended scientific trip to Central and South America. As he roamed, collecting unknown creatures to take back to Europe, he also took readings of magnetic intensity. The weakest field he found was in the town of Micuipampa in northern Peru and on that basis established the value of the global field in that town as one unit of intensity, as Gillian Turner explains. That meant future measurements of intensity could use the Peruvian unit as a reference point. It was a start, and could provide a snapshot of the relative strength of the field for at least a few points on Earth. Von Humboldt began to dream of a worldwide network of magnetic observation stations that would measure declination, inclination, and intensity relative to Peru.

In 1828, he met the German mathematician Carl Friedrich Gauss. Gauss, the son of illiterate parents, learned numbers before he learned speech, it was said. A famous story tells of the precocious three-year-old correcting his father's payroll sums and supplying him with the right answers. Today he is known as the prince of mathematics. Gauss figured out an elegant formula for calculating absolute magnetic intensity and published it in 1832. He also devised an instrument, the first magnetometer. It involved one magnet pulling at right angles to another, allowing him to calculate the strength of the magnet measuring the dip, and thus of the Earth's force. It swept the scientific world, becoming the standard for magnetic observatories. Today, we have accurate measurements of the intensity of the Earth's field going back to 1840. Suddenly, the ability to measure the whole magnetic vector was complete. Not only that, but by 1838, Gauss had proved mathematically that the main part of the Earth's magnetic field was generated within the Earth itself. Finally, Gilbert's bold experimentation from more than two hundred years earlier could be shown to be correct.

In the meantime, von Humboldt pressed on with his determination to establish a global network of magnetic observatories. He was in search of the big picture, and to do it systematically he needed observatories across the hemispheres using standardized instruments and collecting information at the same moment. He enlisted Gauss and many others around the world in the effort—including Tsar Nicholas I of Russia. In 1834 the Göttinger Magnetische Verein (the Magnetic Union of Göttingen) was born, named for the city in Germany where Gauss was based. It was the beginning of what came to be known as the magnetic crusade—the first global scientific collaboration and the predecessor to CERN, the European organization for nuclear research that is examining the characteristics of the tiniest parts of atoms.

Longitude still figured into the urgency. Technically, the problem of longitude had been solved by 1759 when the Yorkshire-born clockmaker John Harrison finished his masterpiece: a handheld watch that could keep time at sea. It was known as Harrison's H4, because it was the fourth and most perfect of his mariner's clocks, the fruit of three decades of work for Harrison and his son.

The longitude problem was so consuming throughout the eighteenth and nineteenth centuries that Parliament set up a Commission of Longitude and passed longitude acts, offering lavish rewards to whoever solved the problem for mariners. The act of 1714, for example, offered as much as £20,000, which is more than £2 million in today's funds. Finding longitude was imperative for the "Safety and Quickness of Voyages, the Preservation of Ships and the Lives of Men," the "Trade of Great Britain," and, not least, "the Honour of [the] Kingdom," the act's authors said. It caused a stampede of theories, most of which were bunk.

But Harrison engineered a clock that could keep time at sea, and time, as any sailor knew, was also distance, which was also longitude, because the Earth rotates 360 degrees in twenty-four hours. While Harrison's fourth clock, remorselessly tested against celestial longitude readings in overseas voyages first to Jamaica in 1761–62 and then to Barbados in 1764, could keep time beautifully, it was not easily replicable. Instruments of the day were rare and expensive. Noodle through any museum in the world that contains compasses or quadrants or sextants or other navigational instruments of that period and feast on intricately engraved, lovingly burnished, precisely made works of art. This was not the era of the cheap knockoff.

The Longitude Commission, ultimately known as the more self-important Board of Longitude, wanted a solution that every steersman could lay his hands on. Harrison eventually won the reward, after years of controversy. But lingering questions remained about whether reading the moon or the sun or the stars to find longitude could prove cheaper and more widely accessible. In fact, there was a vibrant and powerful camp within the European scientific community still arguing that the true solution to longitude lay in the celestial sphere. Even at that time the Astronomer Royal—a job Halley once had—was considered the British Empire's prime expert on longitude. The holder of that position ruled over the Greenwich Observatory, founded in 1675 by Charles II for the express purpose of gathering astronomical data in order to find longitude at sea. Astronomy and longitude were inextricably linked. And that meant paying attention to the Earth's magnetic force.

This is where the Irish soldier and astronomer Gen. Sir Edward

Sabine entered the fray. He came to rule the magnetic crusade with what has been called near-fanaticism. It was the 1830s, "one of the most turbulent periods in the history of British science," as one historian has called it. Britain's Board of Longitude had recently been disbanded. The day-to-day issue of longitude at sea was still unresolved. The British Association for the Advancement of Science was newborn. The gentleman naturalist Charles Darwin was setting out on the voyage around the world in the *Beagle* that would last nearly five years and lead him to his theories of evolution and natural selection. (His captain, Robert FitzRoy, took magnetic dip measurements while Darwin looked at the plants and animals.) Queen Victoria would assume the throne that decade, and the mania for collecting data points would blossom into a national craze. Empiricism was no longer a scientific sin but rather an imperative.

Sabine already had a passion for magnetism. He had sailed twice to the Arctic collecting magnetic dip readings and had spent part of the 1830s conducting the first systematic magnetic survey of the British Isles. When he met von Humboldt in 1836, the passion turned into magnetic fever. Sabine became the obsessed driver behind the campaign to convince the British government and its scientific and naval organizations to fund more observatories and finance analysis of the data. The scientific mission took on a zeal usually reserved for the social and religious crusades of the era, such as antislavery and temperance campaigns. Understanding magnetism graduated from a private enterprise to a fully funded national objective, with the stamp of approval of the admiralty itself.

In part, the fervor was about proving British scientific supremacy. Many of the agitators for the magnetic crusade saw their European rivals as the leaders in magnetic research: the Germans, with von Humboldt and Gauss; the French, with the outstanding Paris Observatory and the Bureau des Longitudes. And there was Britain, with its growing empire and its vaunted naval power, both of which depended on navigation, lagging behind in magnetism. It couldn't be borne.

So Sabine masterminded the establishment of observatories in some of the colonies: Toronto; the tropical island of St. Helena; the Cape of Good Hope; and what is now Hobart, Tasmania. Because he knew that the poles were critical to the magnetic quest—the dip needle pointed either straight up or straight down at the poles, for unknown reasons—he also argued successfully for a magnetic voyage to Antarctica and made sure that Sir John Franklin's expedition in 1845 to complete the Northwest Passage in the Arctic carried a superb supply of the very latest equipment for magnetic readings. As I mentioned in the preface, Franklin's mission ended in disaster. All 129 sailors died. Those who'd survived after two brutal winters stuck in the ice abandoned ship, took to the nearest icy bit of land, King William Island, and resorted to cannibalism before every last man perished. But during that second winter, an elite team appears to have conducted magnetic readings close to the magnetic north pole, a priority for the expedition. Eventually, Sabine would take the international reins of the crusade. He wanted the three magnetic measurements— declination, inclination, and intensity—to be observed hourly and sometimes even more often at each of the stations. Gauss and other researchers were aghast at the torrent of data. Sabine soldiered on, ordering his British-based team of scientists to compile and analyze the global findings from all the observatories.

Taken as a whole, the magnetic crusade was "by far the greatest scientific undertaking the world has ever seen," according to a historian of the day. By 1840 there were more than thirty permanent observatories spanning the globe, including eleven supported by the Russian government; four in Asia financed by the East India Company; six in British colonies supported by the British government; and two at universities in Philadelphia and Cambridge, Massachusetts. The scientific world was determined to crack the code of the Earth's magnetic force, aided by systematic observation of data. It was not enough to know that the force changed, or to be able to calculate its strength. Scientists wanted to understand the laws that governed it.

The great physicist Sir Isaac Newton had revealed the laws governing gravity in 1687 (Halley had helped persuade him to publish them), and the last remaining earthly conundrum was magnetism, the great minds of the day believed. Finding the formula to magnetism would complete what Newton had begun—"a revelation of new cosmical laws—a discovery of the nature and connexion of imponderable forces," as William Vernon Harcourt, a founder of the British Association for the Advancement of Science, declared in 1839. Sabine and his colleagues were seeking what Gilbert had sought two centuries before—a comprehensive new way to understand how the world worked. The timeless secrets of the universe were there, ready to be unlocked, tantalizingly within reach. They wanted the key.

As the crusade wound down toward the end of the 1840s, it was clear to most of the participants that they had not found that key. One British science historian wrote that "it could be argued that scientifically the results were not worth the massive efforts." By the 1850s, the push to figure out the Earth's magnetic force had receded, supplanted in part by the astonishing new biological theories of evolution and natural selection Darwin developed after his tour around the world on the *Beagle*. The question of how the planet worked faded in the face of the far more controversial questions of how life itself had come to be. Ruefully, the physicists and geographers concluded that the magnetic force would never really matter to the everyday functioning of the world. It became a sidelined scientific curiosity that no longer demanded explanation.

Still, there were some advances. Sabine, poring over the data and analyzing it, was able to show a connection between the occurrence of odd dark spots on the sun and transient fluctuations in the intensity of the Earth's magnetic power. It was the first hint that the two might be linked. That link would prove critical to the new generation of scientists who are now trying to predict what the Earth's fickle magnet will do.

The larger benefit of Gauss's union of observatories was that it provided the first in what is now more than 175 continuous years of measurements of the Earth's magnetic field. The first indications, in fact, that the field is growing weaker, which the poles must do before they flip. And while the quest to understand magnetism fell behind in the nineteenth century, forgotten in the exhilaration of new scientific pursuits, it did not remain there for long. A new urgency would soon arise, a modern magnetic crusade, as scientists struggle to understand how the reversal of the poles will affect civilization.

CHAPTER 9

the rock that turned the world upside down

To drive the countryside with Kornprobst was to learn to read the tale of the Earth's tormented dramas and fiery convulsions, its traumatic passage through time. After decades of practice, it came as easily to him as breathing. "Basement! Three hundred million years old!" he declared as he maneuvered down a narrow road, pointing to his right at a fragment of the continental crust. A little farther, gesturing to something that looked precisely the same as everything else around it: "A small volcano from fifteen million years ago. Only its chimney is left." Farther still: "Basalt. Ten thousand years." Two beats later: "Lava. Forty thousand years." Then: "Basalt!"

We were on our way to the outskirts of the tiny village of Pont Farin, perhaps a two-hour drive from Clermont-Ferrand, to see if we could find Brunhes's seam of terracotta. The snow had melted overnight. The damp chill had lifted. The sky was clear cerulean blue and the fields of central France were waking from winter, a few green patches standing out here and there among the tidy rectangular farms. Just weeks later, these fields would be sown with corn, potatoes, and wheat, nurtured by the rich black soil that is the legacy of this area's ancient string of volcanoes. Volcanoes have fed the livelihood of every generation that has settled here, with crops, pastures, thick creams, crumbly cheeses, tannic wines, and eventually industries that relied on stream water made pure by the exquisite filter of the rocks.

More than a century earlier, Brunhes had put out feelers to the road engineers of this rural heart of France, asking them to be on the lookout for a formation of sedimentary terracotta that had been covered over with hot lava from an ancient volcano. Like so many other geophysicists of his day, Brunhes was trying to find rocks that had lost and then gained a magnetic fingerprint after being superheated with lava. According to Melloni's findings, Folgheraiter's conclusions, and Curie's rule, terracotta in that formation ought to have the same magnetic orientation as the material that later spilled on top of it, the lava or basalt, a fine-grained black rock. The larger goal was for physicists to try to track long-term magnetic variations through rock samples taken from different parts of the world, reconstructing the evolution of the Earth's magnetic field over time and trying to work out what could possibly cause it to change.

And then one day, one of Brunhes's friends, a Monsieur Vinay—history has not recorded his first name—an engineer with the road and bridge works administration, told him about a new road he had helped excavate near Pont Farin, or Pontfarein, as it was called then. The construction of the road exposed exactly the configuration Brunhes was looking for: a long seam of terracotta covered with basalt. Brunhes packed his chisels, got on his horse, and set off.

Kornprobst was zooming down the Giscard d'Estaing autoroute, named after the former French president, heading south. He had his maps out. The road cut was going to be tricky to find. Even Pont Farin was not obvious, tucked away on a loop of rural road that Google

Maps barely notices. I could see that he was a little nervous. He speeded up, darted over, passed a string of cars, and then zipped back into the right-hand lane. Another car honked at him angrily and then whipped past. Kornprobst took his right hand from the gearshift, held it up in front of the rearview mirror, palm facing in, and gave a dismissive wave, mouth set in a line of determined nonchalance.

He scanned the landscape. To a geophysicist, coming to this part of France, known as the Cantal, was like reminiscing with a cherished friend over a sumptuous dinner. The autoroute cut through layers of sedimentary rock laid down 30 million years ago during the Oligocene, the epoch when truly archaic animals gave way to some we would recognize today: elephants, pigs, horses, and apes. All that sediment lay on top of a hard granitic basement, forged in heat long ago and then cracked in places and thrust to the surface by the gyrations of the crust itself. Underneath all of it were the remains of magma beds that had fed volcanoes much older than the Chaîne des Puys. Formed many millions of years ago, they had spilled such immense quantities of basalt that it formed a lake of the stuff. Basalt is beloved by geologists because it crystallizes in the mantle and is therefore the closest thing to original lava they ever see. Now the volcanoes, once the greatest Europe had ever seen, were worn down to shadows of their former glory by winds, rain, and time. And the lake of basalt had been transformed into a vast fertile plain.

By 10:45 a.m., we were on picturesque back roads, arriving in Les Ternes, a village of six hundred built in tiers up the side of the road. Tidy stacked stone walls lined its few streets. Kornprobst suggested casually that we stop for a coffee at the restaurant that stood below the town's sixteenth-century castle. Just as casually, he approached two middle-aged men sitting at its splendid bar. One was in blue jeans, the other, camouflage fatigue pants. Did they know Pont Farin?

Mais oui!

Kornprobst listened intently as they described how to get there:

up to the top of the village to the road running along a fenced pasture, and bear to the left. He thanked them, and then couldn't resist. Did they know that a famous physicist discovered just a little way from here that the poles reverse? That north becomes south and south becomes north? That a visitor from overseas—he pointed to me then—was here to write about this very, very famous physicist? The men looked at me, smiled politely, shrugged at Kornprobst, and went back to their drinks.

There were no markings on the road once we found it at the top of the village, just a tiny metal flag on a post naming the road as D57. Again, Kornprobst consulted his map. Yes, yes, we were on the right track, he murmured as we wound our way through the fields. Past more stone fences and low roofs covered with moss. And then, Pont Farin. Proudly, Kornprobst stopped the car and snapped a photograph of me standing in front of the village sign, clutching my notebook.

"Village" may be an overstatement. It had two houses and a stream. But farther on, past the houses, was another yet smaller curve. Kornprobst pulled tentatively onto the bank of the road. Was this it? He hesitated. No. He pulled forward again and then spotted something that looked similar. Here! He was jubilant. He threw open the trunk of the car and retrieved his geologist's toolkit: a compass in a battered red casing and a hammer.

We set off at a clip down the narrow road. It hugged the side of a hill, a sharp cliff off to the left down to the river far below. This was the road Brunhes's friend had helped make more than a hundred years before. I couldn't help but marvel at the fact that it was still there. No village sprawl. No road widenings. No big-box stores or industrial complexes paving over the old site. This part of France has remained largely unchanged for hundreds of years. Birds were chirping merrily. The scent of manure was strong. The sun was finally warm enough to encourage us to shed layers of clothing. We were utterly alone.

Ah! Terracotta! Kornprobst cried, pointing to a sprinkle of brittle red flakes in the pool of a stream trickling down the bank of the hill on our right. We were getting close. The first time he came to find Brunhes's terracotta, he failed, he told me. He had been expecting a quarry. Instead, it was a nearly untouched outcropping covered by the detritus of a century. I was expecting a sign or some small mark of the scientific revolution that had been wrought in this place. There was nothing. A more forgotten corner of France would be hard to imagine. We had been walking for a while by now. He spotted a house just around a curve and stopped short. Too far. We turned back. Now he was even more intent, stooping down to peer at the bank for the telltale red of the terracotta seam and occasionally banging at it with the hammer.

Suddenly, he leapt across the stream and clambered up the side of the hill. It was strewn with plastic oil containers and discarded wine bottles. Young trees were struggling to survive in the thick layer of soil. He leaned down with the hammer and whacked at a piece of moss. A chunk of terracotta rolled out into his hands. His face cracked into the gleeful smile of a child as he handed it to me.

This jagged piece of rock nestled in the palm of my hand had been laid down right here 10 to 15 million years before, undisturbed until this moment. Five million years ago, a volcano erupted, spilling lava over the expanse of terracotta, heating it up to as much as 700 degrees Celsius, well above its Curie point. Recall that terracotta contains a lot of iron. Each iron atom has four unpaired electrons, and they got so hot when the lava covered them that they lost the magnetic orientation they had held for all those millions of years prior. Then, when they cooled, shortly after the volcano erupted, they realigned themselves with the magnetic field's direction and intensity of that particular spot on the Earth at that time. Enough unpaired electrons in the terracotta lined up with the Earth's magnetic field that the rock's magnetic flow locked into the direction it had at that time. In effect, the terracotta became a crust-bound magnet.

But when Brunhes took his samples from this very seam of terracotta back to his laboratory in the Rabanesse tower in Clermont-Ferrand—he had a lot of trouble getting perfectly shaped samples because the terracotta was so fragile—he found that the magnetic dip, fixed by time and heat into the rock to show the direction of the field it had cooled in, pointed in exactly the opposite direction from what he believed to be north. It was the same story with a piece of the basalt he had broken some chisels on as he extracted it from just above the terracotta.

To him, the conclusion was inescapable: When the terracotta heated up and then cooled underneath its layer of lava, the magnetic north pole had been on the opposite side of the Earth from where it was in France in 1905. He published a paper on his findings in 1906. It was a momentous couple of years for the science of electromagnetism. J. J. Thomson won the Nobel Prize for discovering the electron, the first subatomic particle, in 1906. Albert Einstein published his special theory of relativity in 1905, laying the groundwork for the vast electromagnetic infrastructure that is the central nervous system of modern civilization. It would be another fourteen years after Thomson's prize until the proton was discovered in 1920 and a dozen more until the neutron was identified in 1932. And it was even longer, until after the Second World War, before the role of the unpaired spinning electron in the phenomenon of magnetism would be fully understood and the science of the reversals of the Earth's magnetic field became incontrovertible.

But in 1906, the implications of Brunhes's claim that the poles had once been reversed were so staggering as to make it unbelievable to most scientists. They spent decades deriding it and questioning whether the Earth's rocks were a reliable record of magnetic memory. They still hadn't worked out how or why or on what schedule the Earth's magnetic field's declination, dip, and strength changed. The idea that the field could reverse direction would mean that they had critically miscalculated the nature of the force itself. And that they had no clue about what made it function in the first place. Apart from being a revolutionary idea, it was an assault on scientific pride. And it was easy to scoff at because the conclusive proof that would eventually convince the scientific world still lay concealed in the Earth's crust, under the sea and deep within its core.

Brunhes never published on the magnetic field again, although he continued to research it. On Sunday, May 8, 1910, he had just returned to Clermont-Ferrand from a trip to the Cantal, where he had been taking measurements in some mines. The weather was foul, and it was snowing heavily. Despite the snow and his own fatigue, he left the Rabanesse tower near midnight to find out the results of a civic election in a nearby town. Shortly after, police officers making their rounds in the ancient streets near the tower found a man unconscious on the ground. It was Brunhes. They took him home, but he'd had a massive stroke. He died at noon on Tuesday, May 10. He was forty-two.

He did not live to see his grand new observatory, Les Landais, which opened two years later. He did not live to see the vindication of his outrageous theory of reversal, or to discover that the Earth's magnetic direction had flipped not just once but hundreds of times in its history. He would never know that this current magnetic phase of the Earth's history is named after him. The man who traveled by mule and horse did not live to see all the new data from satellites orbiting Earth showing that its field is becoming daily more disturbed, and that part of it in the southern hemisphere has already changed direction. He could not have known that a century after his death scientists would be debating whether the poles are gathering strength to reverse once again. Or that the vast electromagnetic infrastructure humans have built will be in danger when the poles switch places. He could never have foreseen that scientists are struggling to understand what such a reversal could mean for life on Earth, the great spinning magnet that is our world, or the likelihood of it being utter catastrophe for human civilization.

PART II

current

Our physics, therefore, will no longer be a
collection of fragments on motion, on heat, on
air, on light, on electricity, on magnetism, and
who knows what else, but with one system we
shall embrace the world.

—Hans Christian Ørsted, 1803

CHAPTER 10

experiment in copenhagen

The Niels Bohr Institute in Copenhagen, birthplace of modern physics, was festooned with a modern art installation the day I arrived. The building was covered with an array of lights connected to CERN, the European nuclear research laboratory more than a thousand kilometers away, and they blinked to life whenever subatomic particles, ruthlessly torn apart in the magnetic field within its underground Large Hadron Collider, bashed into one another. Which of the LED lights became bright, and how fast and for how long, depended on which particles were hurtling toward one another in CERN's engineered replica of the primordial cosmos. It was, said the artists who conceived it, as if the music of the universe's birth were being transposed into a symphony of light on the building's dun façade.

This stately, red-roofed set of laboratories and offices was built in honor of the Danish physicist Niels Bohr in 1920. Bohr won the Nobel Prize two years later for developing the first simple image of the atom's internal architecture: electrons buzzing around a nucleus that is fashioned of protons and neutrons clinging together. Bohr peered into the heart of matter for the first time, and his institute became a mecca for theoretical physicists from all over the world in the sensitive era between the two world wars when physicists were wrestling with the potential of splitting those nuclei to make the atomic bomb. Even the institute's address, Blegdamsvej 17 (17 Bleaching Pond Road, named after the nineteenth-century Copenhagen laundrymen who wet linen in designated ponds and then let it bleach in the sun), is evocative to the physics community, much as London's 221B Baker Street is for fans of Sherlock Holmes and detective fiction. The institute was named a historic site in 2013 by the European Physical Society for the sheer magnitude of breakthroughs in physics conducted within its walls.

Andrew D. Jackson, a theoretical physicist with a wry sense of humor, met me at the door and welcomed me in. He was unfazed by the legend of the place. Grinning, he gestured upstairs. The bathtub of the German wunderkind physicist Werner Heisenberg, who won the Nobel Prize for his work in founding quantum mechanics, was upstairs. Did I want to see it? Jackson asked me, chuckling, before leading me into his office.

Jackson, whose voice still bears the trace of his New Jersey upbringing, normally works with some of the more obscure—and sometimes purely conceptual—pieces of the atom that did not figure into Bohr's original image. He has written about mysterious subatomic bits such as skyrmions and about wave packets known as solitons. But a few years ago, one of those curious coincidences that skitter through science propelled Jackson and his Danish wife, Karen Jelved, a scholar of English literature, into the sphere of the nineteenth-century Danish scientist Hans Christian Ørsted, whose work provided a piece of the electromagnetic puzzle that irrevocably changed physics.

In the summer of 1993, the Harvard University historian of science Gerald Holton was in Demark visiting a friend who owned a medieval castle south of Copenhagen. On its library shelves, which had been filled with books purchased by the meter, Holton came across a rare first edition of a work by Ørsted, who died in 1851. As it turned out, Holton had been awarded the prestigious Ørsted Medal for teaching physics in 1980, and as a result had given a talk decrying the invisibility of Ørsted. For one thing, Ørsted wrote mainly in Danish and German and few of his works were translated into English. For another, by the time he died, Ørsted's ideas were wildly unfashionable. Yet his work tracked the tectonic shift from Romanticism to modernism in science during the nineteenth century. How fitting it would be, therefore, to introduce the world to Ørsted's immensely important works, Holton mused to his host. She, in turn, was a friend of the physicist and science historian Abraham Pais, a colleague of Albert Einstein who had once been Bohr's assistant, and she told him about her dreams for broadcasting Ørsted's legacy. Pais and Jackson had lunch together every day, and Pais knew that Jackson and Jelved were looking for a new project that could make use of their skills in both language and science. Pais talked to Jackson. And that, as they say, was that.

Jackson and Jelved have become known as the English voice of Ørsted, publishing his works on science, literature, poetry, and philosophy. Few understand him better. ("It's my experience that most good scientists are romantics, and it's pretty lonely," Jackson told me as he tried to help me understand Ørsted's life.) They helped translate a key volume of his selected scientific works, as well as plowing painstakingly through the letters he wrote home during eight journeys within Europe in the first half of the nineteenth century. Ørsted wrote his letters in a Gothic script, using a quill pen and paper so punishingly thin that the ink soaked through. Worse, he frequently underlined words for emphasis. At one point, worried about the money he was spending on postage, he decided to start writing just as much, only half as big, Jackson told me, groaning slightly.

Not only that, but Ørsted's daughter, Mathilde, who published a short, heavily edited version of his letters in 1870, eradicated evidence of her father's aborted first engagement. In 1801, Ørsted had pledged to marry his mentor's household employee, Sophie Probsthein. Ørsted finally married Mathilde's mother, Inger Birgitte Ballum, known as Gitte, in 1814. The attempt to blot out evidence of the relationship succeeded until Jackson and Jelved pieced together evidence of the hidden love affair.

Ørsted's travel and professional life coincided with the heady decades of the first half of the nineteenth century, which have since been called the Danish golden age. In the few years straddling the beginning of that century, Copenhagen, then home to about one hundred thousand, had suffered through two widespread fires and two devastating bombardments by British forces. The city needed to be rebuilt and the process spawned a creative surge that spanned architecture, literature, music, visual arts, and finally science. Hans Christian Andersen, who wrote the beloved fairy tales, was among the most famous figures of the age and was a close friend of Ørsted.

Ørsted took the golden age at a canter, and the letters show him using his journeys and growing fame to try to raise Denmark's status within Europe. He was an inveterate name dropper, Jackson confided, and he met nearly everyone of any consequence. One of the highlights of his life was meeting his idol, Sir Walter Scott, on July 4, 1823, in Edinburgh.

His travels also introduced him to some of the magnetic ideas floating around Europe at that time. During the 1823 trip abroad Ørsted made measurements of the Earth's magnetic field and, later, on his final journey in 1846, traveled by steamship down the River Thames from London to Greenwich—under the arches of Waterloo Bridge, Blackfriars Bridge, and London Bridge, he recounted with obvious delight in a letter home—to visit its famous magnetic observatory.

In 1827, he journeyed to Altona, near Hamburg, Germany, then under Danish control, to take part in a meeting of the finest magnetic minds in the world. The peripatetic Alexander von Humboldt, who had set up the first system of relative measurements of a magnetic field's intensity based on readings from a village in Peru, was there. So was the brilliant mathematician Carl Friedrich Gauss, who later worked out how to measure the field's absolute intensity. Along with Ørsted and others, they proposed the establishment of the first global network of magnetic measurements. The meeting resulted in the formation of the Magnetic Union of Göttingen, which launched in 1834 and was the first international scientific collaboration and the forerunner of CERN.

It wasn't just measurements the Altona group wanted. They also pushed for a plan to finally pinpoint the location of the Earth's magnetic north pole. Just four years later, on June 1, 1831, it happened. The Arctic explorer James Clark Ross, stuck with his small ship and crew in the ice, discovered it near the very northern tip of what is now mainland Canada using a magnetic needle suspended from a silk thread. He built a cairn out of stones to mark the spot, raised the British flag, and called the area British. It was a rich piece of luck that he discovered it at all. The pole had been on a lengthy trip south and was near its most southerly point in centuries. It has been heading north nearly ever since, now veering away from Canadian territory and into Russia.

Ørsted is not invisible in the way that France's Bernard Brunhes is. Rather, it's the breadth of his achievements that has slipped from public memory. He is a cherished figure in science for a single experiment he did in 1820, not for his decades of scientific exploration. On the basis of that one experiment, conducted in public, he is considered Denmark's scientific genius of the nineteenth century just as Bohr is the genius of the twentieth. Today, a park in Copenhagen is named after Ørsted and his famous younger brother Anders Sandøe (who was an architect of the Danish constitution); a university he founded is the international home base for a set of satellites tracking the Earth's magnetic field; a satellite named after him is still in the skies; the Ørsted law is a centerpiece of the physics of steady electrical currents and the Ørsted unit of measurement is part of the scientific description of magnetism; and physics prizes and a coveted fellowship are awarded in his name even now.

His grand finding? In 1820, after having thought about it for the better part of two decades—and against the reigning scientific doctrine of the day—he conducted an experiment showing that magnetism and electricity are physically connected. The find galvanized research across Europe, leading the physicist Michael Faraday to create the prototype of an electrical generator the following decade. That inadvertently sparked the Second Industrial Revolution. (There is a tale—likely apocryphal—that, when asked by the Chancellor of the Exchequer how electricity could possibly be practical, Faraday quipped: "One day, sir, you may tax it!") Over time, Faraday's findings led others to develop the mathematic equations that describe electromagnetic theory. Ørsted's discovery was one of those rare moments in scientific history that change everything that comes after. I was in Copenhagen hoping that Jackson would help me understand why.

CHAPTER 11

a very intimate relationship

For almost the entire time that scientists have been examining electricity and magnetism, they have believed that the two were different. But actually, magnetic and electrical phenomena are not just connected, they are facets of the same thing. That's why physicists now call this fundamental force of the universe the electromagnetic force. "Magnetism and electricity are not independent things . . . they should always be taken together as one complete electromagnetic field," Richard Feynman said. The two, he said, waxing lyrical, have a "very intimate relationship."

To understand the electrical force, we have to go back to electrons and protons. The electron is negative. The proton is positive. These electric charges are the sources of the electrical field. Like the magnetic and other fields, the electrical field is the stuff of the universe, stretching out through it in fluidlike lines that can move in peculiar ways. Electric and magnetic field lines tend to go hand in hand. But there are a few differences between the two fields. While magnetic field lines run in unending loops, electrical field lines can end. And while electric charges can exist as solo particles that are either negative or positive—like electrons and protons—every magnet known in nature has two poles, north and south, just as Petrus Peregrinus discovered in the thirteenth century. No matter how small a magnet gets, those two poles are always present. (Scientists keep looking for a magnetic monopole but have not yet found one.) That means there are no independent magnetic charges.

So where does the magnetic field come from, if not from magnetic charges? Here's where things get a little more complicated than the unpaired spinning electron. It turns out that the magnetic field depends on electrical charges. While the Earth is the source for the Earth's gravitational field and electrically charged particles are the source for the electrical field, it's the electrically charged particles themselves that create the magnetic field, but only when they are moving. In other words, a stationary charged particle makes an electrical field but not a magnetic one. A moving charged particle makes an electrical field and an electrical current, which makes a magnetic field. That can mean a bunch of moving charged particles in a current, or it can be the spin of an electron within an atom. You can take the idea down to the scale of a single atom of iron. Its negatively charged unpaired spinning electrons are creating a tiny circulating electrical current. That means the atom itself is also creating a tiny magnetic field. If you put enough of these atoms together so that the tiny magnetic fields arrange themselves to amplify one another instead of canceling one another out, you get a magnetic substance. In effect, as Feynman said, all magnetism is produced from currents of one sort or another.

Albert Einstein realized that what constitutes "movement" here depends on one's frame of reference. If you are at rest with respect to an electrical charge, you will see an electrical field. If you are moving with respect to the same electrical charge, you will see a moving charge, which is producing an electrical current as well as a magnetic field. The same is true when you are stationary with respect to an electrical charge that is moving. It's all about perspective. It's all, as Einstein would say, relative.

The journey to entwine electrical and magnetic forces culminated with Einstein. But it has scampered across thousands of years, winding through myth, dogma, experimentation, and, finally, mathematics. The fact that these phenomena are facets of each other came as a surprise. Consider this: The name "electromagnetic," one of the many words that Ørsted coined, contains within it William Gilbert's cranky christening five hundred years ago of the word "electricity" from the Greek word for "amber" as well as Homer's telling of the tale of the ancient hero-king Magnes nearly three thousand years ago, patched onto the evolution of those ideas in the centuries since. Were we to magically erase all that rich history and metaphor embedded in the current label and name the electromagnetic force anew, knowing what we know about physics today, we would give it a label that clearly indicates that magnetism and electricity are the same thing.

The electromagnetic force is one foundation on which the whole universe rests, at play in each single small piece of each atom. The electromagnetic field can manifest itself as waves, or vibrations, that can be any length. We see some of the tiny waves in the form of light and color, which means, by definition, that light is also electromagnetic. By a pleasing symmetry of nature, the charges all seem to balance one another out most of the time, making the universe electromagnetically neutral. Most of the time, we aren't even aware of the electromagnetic force that is so powerfully at work.

So, the electrical force is produced by charged particles. Electricity, on the other hand, is electrons in motion. The earliest forms of electricity that scientists became aware of were what today we would call static electricity. A spark is static electricity, and so is lightning and so is the emanation from rubbed amber that attracts a piece of fluff, the phenomenon that led Gilbert to name electricity in the first place.

You're making static electricity when you rub a balloon on your hair. The balloon steals a few electrons temporarily from your hair, making the balloon slightly negatively charged and your hair slightly positively charged. If you hold the balloon overtop your head, your hair will fly up to meet it. The hair's positive charge wants to be reunited with the balloon's negative, and the force between them is strong enough to lift your hair. Eventually, the electrons drift away from the balloon and the hair falls back down. The rubber in the balloon is called an insulator because it doesn't easily conduct electrical charges. Insulators used to be called "anti-electrics," and include other things such as glass and wood and plastic. When insulators capture extra electrons, they store them, like the balloon does, rather than pushing them somewhere else. Insulators can also isolate pockets of opposite charges from one another.

The revelation for scientists in the late eighteenth and early nineteenth centuries was that they could make electricity flow. At that time, they believed that electricity was a fluid that ran through wires and so they called it "current" electricity, as if it were a running river. Today, we say that electrical current is the one that moves from the socket through the wires into your lamp. What makes that happen? The electromagnetic field can be harnessed to push electrons through substances known as conductors and make them travel from one place to another. The human body is a conductor. Others include metals that have at least one unpaired electron in an outermost filled orbital, like copper, which is often used in electrical wiring. The old incandescent bulbs used to have a metallic tungsten strip in them where the electrons collected and heated up the metal, producing heat and light. New LED bulbs, like the ones in the art display at the Niels Bohr Institute, shine because electrons are forced into shedding some of their energy in the form of tiny units of light called photons, which are extremely short electromagnetic waves.

A key to all of this is that while the electromagnetic force is fundamental to the universe and will live as long as the universe, the process for harnessing that force into making a current is only about two hundred years old. And forcing immense amounts of current into a vast, interconnected transmission system, like the modern electrical infrastructure we rely on, is only about one hundred years old. As a society, we devote significant amounts of time, thought, and money into keeping these systems going. But as the planet's magnetic field undergoes its restless contortions inside the Earth's core, the transmission systems themselves are put at the kind of risk no one imagined when they were created. Under certain circumstances that scientists are just beginning to track, the planet's human-built electrical transmission system could be switched off.

CHAPTER 12

jars full of lightning

While the bid to understand magnetism was an impassioned quest over centuries, fraught with theological risks and potentially stupendous financial rewards, the study of electricity was, by contrast, slacktwisted until the middle of the eighteenth century. Even then, electricity did not question the position of the planets or the sun or the age of the Earth. It did not endow the planet with its own soul or try to wrest the Bible from its perch of authority. It was "cosmologically neutral."

Early philosophers, including the same canny Thales of Miletus who cornered the ancient Greek market in olive presses and looked at the power of the magnet, also examined electricity. Thales is said to be the first to realize that if you rubbed amber, which is a honey-colored fossilized tree resin, it could attract pieces of chaff—a phenomenon he noticed because Greek women of the seventh century BCE occasionally spun wool with precious amber spindles. Some of those spindles can still be found in museum collections today.

But to the ancients and most medieval researchers, electrical draws were, seemingly, even more impermanent than the pull of the

lodestone. Under certain circumstances, if you rubbed amber, it created sparks, but not always. Occasionally, pieces of feather or chaff
would temporarily stick to the rubbed surface of amber. But damp
or rainy days scotched any deliberate attempt to make sparks or attract chaff. The sparks and the sticking chaff are static electricity, the
result of electrons being shaken free from their orbitals, producing
slight electric charges and temporarily lodging in other orbitals.
When a material such as amber is damp or the air surrounding it is,
the water acts as a conductor for the electrons, ferrying them away
and preventing them from building up into a spark. In the minds of
the early researchers, electricity was not a key to understanding the
universe. It was a mildly interesting, fleeting, largely inscrutable curiosity. There was certainly no electric crusade.

It took William Gilbert, intemperate physician to Elizabeth I, to
show a sustained experimental interest in electricity. As he was doing research for his treatise *De Magnete*, published in 1600, in which
he explained his shocking conclusion that the Earth's magnetic
power rested deep inside its core, he also looked at amber. He discovered that not only amber but also a range of other substances, including jet and diamond, could be made to attract chaff when they were
rubbed. He named the phenomenon "electricity" and then dismissed
it as inferior to magnetism, the grand force that he believed, wrongly,
kept the Earth in its daily and yearly rotation.

Even Isaac Newton, the towering Enlightenment physicist who,
in 1687, published his mathematical description of gravity, one of the
other four fundamental forces of the universe, didn't get very far
with static electricity. It fascinated him, though. He ran numerous
experiments in the late seventeenth century to try to understand it.
In a note on December 7, 1675, to the new Royal Society of London
for Improving Natural Knowledge (now simply the Royal Society),
he described in confusing detail how to make very thin triangular
bits of paper dance underneath a round piece of glass rubbed and set

overtop a brass ring. The instructions failed. The fellows of the society wrote back to him asking for further directions and finally got the show to work using stiff boar bristles to rub the glass.

As the modern American historian of science J. L. Heilbron explains, Newton's note is remarkable for demonstrating how little even the most accomplished natural philosophers of the day knew about electricity. Even the era's most eminent mathematician could make neither heads nor tails of why static electricity crackled and danced.

A series of experiments begun in 1733 by Charles François de Cisternay du Fay, the wealthy scion of a French military family, was the first systematic attempt to compile findings on electricity from all over Europe. Up until then, the experiments showed scattered, inconsistent results. Du Fay wanted to put things in order. He wanted rules.

His first finding was that everything, except fluids and things too soft to rub, could be made electrical by friction, or excitation, as he called it. It means every substance can produce static electricity if you rub it the right way and if it is dry. And how did that electrical "virtue," or static electricity, get transferred to other substances? Both by touch and by proximity, du Fay discovered. But he added a large caveat: The substance receiving the electrical spark had to be laid on something that did not conduct electricity—an antielectric or insulator. This became known as the rule of du Fay and was followed assiduously for more than a decade.

Du Fay's experiments also convinced him that there are two types of static electricity. We would say that there are positive and negative charges, like the balloon you rub on your hair that is negatively charged because it has picked up electrons and your hair that is positive because it has lost them. But du Fay came to believe, wrongly, that certain substances could have either one type of electricity or the other, but not both.

By the middle of the eighteenth century, electricity was no longer the poor relation of scientific technology. It had come into its own.

Two discoveries made all the difference. The first, discovered independently by at least two experimenters, and named after a Dutch university town, was the first condenser, also known as a capacitor, which temporarily stored static electricity in a glass jar. The second was by the American diplomat and scientist Benjamin Franklin, who fished for electricity in the skies with his kite.

Researchers into electricity, who called themselves "electricians," not only began to catch a glimpse of a future where electricity might be made to do things, but now they also knew that it was somehow linked to the majestic drama of the stormy skies. "Forty years ago, when one knew nothing about electricity but its simplest effects, when it was regarded as an unimportant property of a few substances, who would have believed that it could have any connection with one of the greatest and most considerable phenomena in Nature, thunder and lightning?" said Samuel Klingenstierna, a professor of physics at Uppsala University in Sweden in 1755.

The first big advance came in the 1740s. For years, experimentalists had been able to produce shocks of static electricity through friction, following du Fay's findings. They experimented with making stronger and stronger shocks. While some of the electricians believed that they knew everything there was to know about electricity and that there was no need for further research, others wondered whether electricity was a fluid that they might be able to imprison and transport in a jar. Many of the electricians' contemporaries considered the goal ludicrous, as absurd as boxing a light beam inside a soap bubble, as a later historian of electricity put it.

But then in January 1746, the legendary Dutch physicist Pieter van Musschenbroek, a professor of philosophy at the University of Leyden who had turned down royal sums to teach several of Europe's science-hungry kings, made a breakthrough. He was repeating an experiment made by a lawyer who was an amateur scientist and who had been fiddling around in van Musschenbroek's laboratory. This amateur didn't know about the rule of du Fay that insisted that the substance to be charged up had to be set on an insulating material. So he held a water-filled jar in his hand, electrified it with static electricity, touched the wire carrying the charge, and got a spark.

Two days later, van Musschenbroek repeated the experiment. He hooked a thin glass globe to a metal gun barrel suspended by silk threads. One assistant used a contraption to turn the globe rapidly while another steadied it, sending the electrical charge down the length of the gun barrel. Attached to the end of the gun barrel was a brass wire, which entered a jar partly filled with water. Van Musschenbroek held the jar in his right hand and then tried to draw sparks from the wire with his left. The immense voltage of the static electrical force jumped into his right hand. His whole body shook as if hit by lightning.

The charged electrons created by friction between the globe and the gunmetal rushed across the unpaired electrons of the metal and the brass wire and collected inside the glass, trapped there by the fact that the glass could not conduct the electrons further. The wire became one half of an electrode, and van Musschenbroek's highly conductive hand became the other. Each half of the pair was equally charged until van Musschenbroek touched the wire with his other hand and the charges from the wire rushed to their opposites in van Musschenbroek, carrying immense voltage. The process carried the risk of electrocution, and in fact the instrument, which became known as the Leyden jar after the city where van Musschenbroek lived, was used experimentally to kill animals.

Van Musschenbroek wrote up the experiment in Latin in a letter to the French scientist René Antoine Ferchault de Réaumur. He was still trembling with fear: "I wish to inform you of a new, but terrible experiment, which I advise you on no account to personally attempt," he wrote, adding that he wouldn't repeat the dreadful thing even if he were to be given the whole kingdom of France. "In a word, I believed I was done for," he wrote.

More important, van Musschenbroek didn't understand what he had done. The experiment had defied the rule of du Fay because there was no insulating material under the glass globe. It made no sense to him. "I've found out so much about electricity that I've reached the point where I understand nothing and can explain nothing," he wrote to Réaumur.

The Leyden jar was a revelation, both for science and for high-society entertainment. Future scientific refinements replaced the jar's water with a lead lining, inside and out, creating the two sides of the electrode. And as long as one didn't inadvertently touch the wire going in with anything that could conduct the electrical charge, the electricity could stay inside the jar for hours or even a few days and be released later. Not only that, but the "electricians" soon realized that they could connect one Leyden jar to another and one more again in order to make the shock bigger. It was like a rather cumbersome, short-life prototype battery, differing from modern batteries in that the electricity came from friction rather than chemical reaction.

The jars transfixed the eighteenth-century Enlightenment establishment. As the Cambridge University historian of science Patricia Fara explains, making electricity became an international obsession. All of a sudden, people felt that they could control the spark of life. They had power. It was intoxicating.

It was also dangerous. Citizens and researchers who tried replicating van Musschenbroek's experiments or who offered themselves as subjects to be experimented on reported nosebleeds, passing paralysis, weakness, and dizziness, the result of what today we recognize as high-voltage shock. "I found great Convulsions by it in my Body," wrote the Leipzig classics professor Johann Winkler. "It put my Blood into great Agitation; so that I was afraid of an ardent

Fever; and was obliged to use refrigerating Medicines. I felt a Heaviness in my Head, as if I had a Stone lying upon it. It gave me twice a Bleeding at my Nose, to which I am not inclined."

Still, a conundrum presented itself. Was the electricity—or fire—that men and women could make by friction the same as that made by nature? Lightning, for example, seemed similar to the sparks produced from the Leyden jar. But was it? Or were they two utterly separate entities? The American businessman, intellectual, and scientist Benjamin Franklin set himself the task of finding out.

Franklin is best known today for his role in helping to draft the American Declaration of Independence, his diplomatic efforts on behalf of the British colony of Pennsylvania, and his many inventions, including the stove that still bears his name. But he was also an internationally celebrated, self-taught "electrician" who devised ingenious experiments and made seminal findings throughout his long life. He was awarded the Copley Medal in 1753—the highest scientific award of his day and the equivalent of today's Nobel Prize—"on account of his curious Experiments and Observations on Electricity."

He became fascinated with electricity in 1745, when an American scientist friend sent him a glass wand for experiments from London and breathlessly revealed in a letter that all Europe was agog at parlor-room demonstrations of the new electrical charges. Truly, his friend wrote, they were living in an "age of wonders." Franklin enthusiastically taught himself to perform the demonstrations in his home before throngs of visitors, and then taught them to a neighbor, whom he encouraged to hit the lecture circuit with the electrical oddities. It smacked of a carnival, judging from quotes Fara cites from posters advertising the neighbor's lectures. "A curious Machine acting by means of the Electric Fire, and playing [a] Variety of Tunes on eight musical Bells," reads one. "A Battery of eleven Guns discharged by a Spark, after it has passed through ten Foot [sic] of Water," reads another.

For Franklin, electricity was far more than an entertainment. He ran a spate of careful experiments to explore what it could do. At one point, he methodically disassembled the Leyden jar to discern which part of it held the electrical charge or "virtue," as Joseph Priestley, the eighteenth-century theologian, scientist, educator, and historian of electricity explained. (It was the glass, which served as the insulator.) Like du Fay, Franklin also determined that everything is inherently electric and that electricity has both negatives and positives that have the urge to balance out. He said that substances can be forced to go out of balance so they will have a charge to give off. He also asserted that electrical charge can be neither created nor destroyed; it can only move around. These were remarkably astute observations that hold up today, even if Franklin and du Fay could not express them in terms of moving electrons. The same spirit of logical deduction based on observation led the naturalist Charles Darwin to figure out several decades later that species evolved and adapted to their environments, even before Gregor Johann Mendel published his findings about genes, serving up a fundamental insight into the mechanism for how life forms have changed.

And then there was lightning, the experiment Franklin is most remembered for. It was not as random as it has sometimes been depicted, but rather the stepwise continuation of electrical research he had been conducting for years. His aim was to determine whether lightning was the same thing as the electricity that gathered in the Leyden jar from friction. So, on a stormy day in Philadelphia in June 1752, he made a kite with a silk sail, a wooden spine and spar, plus a strong bridle. To the spine, he attached a wire. To the end of the kite's hemp line, he attached a metal key wrapped in silk, which in turn was attached to a Leyden jar. Lightning lit up the sky, some electrical charge hit the wire (not likely a full strike of lightning, which would have killed him), ran through the line, passed through the key, and filled the Leyden jar. It was indistinguishable from any other electrical spark that the jar had contained. Franklin had drawn fire from the heavens and shown that it was the same as the human-made spark. He was jubilant.

The physics of lightning is still being explored today. But at base, lightning is a long spark of static electricity. As hail, ice, and super-cooled water droplets bang around within a storm cloud, they shake loose electrons. The electrons gather around low-hanging hail, creating a negative mass toward the bottom of the clouds. The positively charged ice crystals move up toward the top of the clouds. When the negative charge builds up enough, a long line of static electricity cracks toward the Earth or toward another cloud, searching for a positive charge. As the electricity moves, its heat makes the air flash with light and expand abruptly, causing the sight of lightning and the sound of thunder.

The Philadelphia adventure was the second of Franklin's lightning experiments to succeed. The first, much less hallowed, happened a month earlier in France. Two French researchers set up a metal pole fixed to a church steeple, after Franklin's instructions, and then waited for a strike. When the lightning struck, their assistants valiantly poked the pole with a brass wire insulated by a glass handle. They drew sparks, just like the sparks in the Leyden jar. It was a risky proposition and other experimenters trying to chase lightning in this way were injured and at least one fatally electrocuted.

Franklin used the findings from the experiments to lobby for metal lightning rods to be permanently installed in high buildings—especially the towering churches—and linked by a metal cable or wire to the ground. He reasoned, correctly, that such a metal rod would collect the lightning charge and ferry it to the ground, where it would dissipate harmlessly. (The same idea is behind having a "ground" in the modern electrical plug.) It was a bid to protect buildings and people from both a lightning strike and the fire such a strike could cause. It was controversial, especially in France. Lightning, it was said, was the punishment of a miffed God. Diverting the lightning was subverting God's will.

Even though electricity couldn't do much yet, the sheer spectacle of it, and the fact that it was linked to the power of the storm, enthralled people. Leyden jars and spark-making machines flourished, furnishing the diversions of the age. At the British court, dancing was abandoned in favor of "electrical entertainment." In France, the electrician Jean Nollet, a disciple of du Fay's, delighted King Louis XV by sending a charge through 180 soldiers, making them leap in near unison. British hosts with questionable taste electrified metal cutlery for the kick of seeing their guests jolted. The electrician Stephen Gray prompted gasps from onlookers with his "hanging boy" trick, in which he suspended a schoolchild, passed an electric spark through him, and caused feathers to fly toward him. A favorite joke was the electrified painting of the king, primed to jolt any republican who happened to flick fingers at his crown.

But while all these shocking tricks may have been amusing, the idea that electricity could be routinely made and delivered to homes and businesses, that it could replace the muscle work of humans and horses, that it could shape a different sort of civilization, was unimaginable. Still more remote was the idea that these sparks that filled a jar could be related to magnets or that either could be in any way connected to either the inner workings of atoms or the contortions of the liquid metal in the Earth's core.

the apothecary's son

Hans Christian Ørsted didn't believe in atoms. In fact, he spent a great deal of his life's work trying to dispute any idea of "atomism," as it was called. He considered atoms devoid of life. They simply could not be what formed the glories of nature that he saw around him. And not just nature. To him, nature and the human spirit were interwoven. Together, the two were always on the move, interacting, shaping each other. They were a dynamic representation of the mind of God, not tiny bits of dead things.

Ørsted came to these beliefs through the teachings of the German philosopher Immanuel Kant, whose ideas helped spark Europe's love affair with Romanticism. Ørsted began to explore Kant's ideas at university and became hooked. They helped him determine what problems he would work on and shaped his interpretation of his results. For good or for ill, Ørsted was using his experiments to prove some of the ideas that Kant espoused.

He wasn't the only scientist of his day to fall under Kant's spell. Kant, who died in 1804, was one of the last philosophers to have had such a profound effect on scientific thought and practice, Andrew D.

Jackson explained to me, relaxing in front of one of the computers in his office at the Niels Bohr International Academy, an independent center of excellence housed at the Institute. His hands were clasped across the front of his navy-blue shirt, eyes dancing with merriment. Jackson helped build this academy into one of the best modern theoretical physics institutes in Europe. ("What we've shown is that bright people want to work with other bright people and our job is to let them do what they want and stand on the sidelines and cheer," he said.) Trained at Princeton University, Jackson got his doctorate in experimental nuclear physics and then taught theoretical physics at the State University of New York at Stony Brook until he moved to Copenhagen in the mid-1990s. He studied shoulder to shoulder with some of the last century's great physicists, including Kip Thorne and the late Robert Dicke, both renowned for their work on gravitational theory and waves. ("I've known them for fifty years," Jackson said.)

Like some of those colleagues, and like Ørsted, Jackson was a broad-gauge intellectual. To sit and talk with him for several hours was to go on an elegant, discursive tour with a philosopher-scientist winding through the history of scientific thought, cultural history, and literature, punctuated with pointed asides on many of the major figures in each of the tales. ("Ritter was barking mad," Jackson confided at one point, referring to Johann Wilhelm Ritter, a German physicist who died in 1810 in his thirties and who was an early influence on Ørsted.) His wide-ranging interests led him to become chairman of the Niels Bohr archives. As a result, he and Karen Jelved translated into Danish the much-traveled stage play *Copenhagen* by the British playwright Michael Frayn, the account of a meeting in 1941 between Bohr and Heisenberg on the role of atomic weapons in the Second World War.

But while Jackson was a well-connected polymath, his subject, Ørsted, was a creature peculiar to the Romantic era. At that time, many thought of science not as a discrete branch of learning but as one thread in the broad tapestry of a sound theological education. Ørsted referred to his scientific work as his "literary career" and called it a form of religious worship. His scientific thinking depended not on atoms or particles, but on the Kantian idea that matter relied on two fundamental forces: attraction and repulsion. Attraction brought matter together and repulsion kept matter from collapsing in on itself when it came together. To Kant, everything anyone observed could be traced back to these two Ur-forces. It was spookily reminiscent of our modern understanding of the four fundamental forces of the universe: gravity, strong and weak nuclear interactions, and electromagnetism. In a way, Kant was not so far wrong, Jackson said, shrugging.

Because Ørsted was such a devout Kantian, he concluded that all the forces observed in nature were derived from these two fundamental push-pull forces. That meant he believed he should be able to find connections and interactions among all the forces he could observe. Not just electricity and magnetism, but also light and heat, motion and air. All facets of nature were somehow bound together; all facets of nature manifested a higher order. It was akin to pantheism, Jackson explained. Since Ørsted believed he knew why they were all related, his goal was to find out how, to develop a unified theory that would explain it all. Because of this overriding philosophy of nature, as well as a belief in experimentation, Ørsted was seen as a progressive, even mildly revolutionary figure in European science at the beginning of the nineteenth century, an important participant in Denmark's golden age. By the end of his life, as he retained his philosophical beliefs while the world moved on, he was regarded as hopelessly behind the times.

Ørsted's interest in science began with chemistry. Modern chemists teach us that chemical reactions are atoms rearranging themselves into new combinations to make new substances, and so it seems odd now to think that an avowed anti-atomist was a chemist.

But to Ørsted, chemical reactions were somewhat hazily defined disturbances of the inner equilibrium of the forces, followed by the restoration of equilibrium.

His passion for chemistry began in the laboratory of his parents' apothecary shop in the town of Rudkøbing on the island of Langeland—or Long Island—in southern Denmark. Hans Christian was making pharmaceuticals by the time he was eleven, under his father's tutelage. Jackson and Jelved did a bicycle tour of Langeland in the summer of 2015, the first time they had seen where Ørsted was born. Jackson leapt to his keyboard, scrolling through tidy piles of files to find photographs from the trip to show me. There was the apothecary's shop, which doubled as an inn when Ørsted's parents ran it. Inside, the brass vessels were still gleaming brightly, still sporting tidy labels, primed to be filled with mysterious concoctions.

It was such a busy place in the late 1700s that Hans Christian and his younger brother Anders Sandøe were sent down the street during the day to the care of a German wig maker and his Danish wife. Jackson had pictures of their home too—a modest building with a sloping roof, still in decent repair. In those days, the nation's time was set according to a replica of John Harrison's famous fourth clock, H4, which had solved the problem of longitude. A man whose job it was to be the nation's timekeeper would travel all over the country with it and people would set their clocks to "Danish normal time." That time remained the country's own until the German Army invaded in 1940, at which time Denmark's time became Germany's. As a result, Danes still say they live on borrowed time, Jackson joked.

A larger-than-life-sized statue of Ørsted stands in Rudkøbing's immaculate square, hands clasped primly in front of his rather portly figure, frock coat to his knees, waistcoat carefully buttoned. In Ørsted's day, the town was too small to have a school, and the brothers picked up what education they could from their parents and the wig maker's family. The boys were multilingual students and voracious learners.

What one learned, he taught the other. Despite being out of the Danish mainstream and despite the ad hoc nature of their education, both became top students at the University of Copenhagen.

Anders Sandøe immersed himself in the law, eventually becoming a Danish prime minister and a famous jurist. Hans Christian devoted himself to chemistry and pharmacy, graduating in 1797. But to his chagrin, chemistry was a degraded discipline at that time. In fact, Kant had deemed it unscientific, just a mechanical process without what he called the dash of intuition or the logic of self-evident truths he demanded in a real science. Part of Ørsted's mission in life was to have chemistry recognized as a legitimate field of research in its own right, not inferior to either physics or medicine.

Two feuding Italian experimentalists helped improve chemistry's standing by discovering what was thought of as "chemical" electricity. The first was the obstetrician Luigi Galvani, whose name lives on in the English verb "to galvanize." The second was Alessandro Volta, whose name and findings are reproduced in the words "volt," "voltage," and "voltaic."

Galvani, who died in 1798, trained in Bologna as a surgeon and anatomist. He was fascinated by how the body came to be infused with life. What would animate it? Like others of his era, he suspected the new phenomenon of sparking electricity. Did its sizzle mean that electricity itself was alive? Could it raise creatures, Lazarus-like, from the dead? Researchers of the day began conducting electrical experiments on dead animals and even people, trying to shock them back into the land of the living.

The novelist Mary Shelley, who was well read in the scientific fixations of her era, captured the fascination with electrical resurrection in her 1818 book *Frankenstein, or The Modern Prometheus*. The subtitle refers to the Greek myth of the immortal titan who steals fire from the god Zeus and gives it to humanity. Prometheus meets a grisly fate; he is chained to a rock, destined to have his liver eaten out each day by an eagle, only to have it regrow each night. In Shelley's novel, the crazed Dr. Victor Frankenstein pieces together a towering man from flesh and bone scavenged from slaughterhouses, dissecting rooms, and charnel houses, disturbing "with profane fingers, the tremendous secrets of the human frame." Finally, "with anxiety that almost amounted to agony," the doctor "collected the instruments of life" around him in order to "infuse a spark of being into the lifeless thing" that lay at his feet.

To the contemporary reader, this "spark" would have been understood to come from instruments to harness static electricity. The experiment worked. Frankenstein's monster came to life: "It breathed hard, and a convulsive motion agitated its limbs." Although, God-like, Frankenstein manages to create a living being, he loathes his creation. It destroys him and kills most of the people he loves, including his wife on their bridal night. Sick of the bloodshed of others, the monster goes off alone to perish in its turn. The novel, considered one of the first works of science fiction in English, reads as a morality tale. It's a repudiation of the idea that humans, or their Promethean electricity-fire, can supplant God.

Galvani's experiments were not about bringing animals back to life but instead about the mysteries of the body's nervous system and brain. At that time, researchers suspected that man-made static electricity and lightning were the same, thanks to Benjamin Franklin's kite experiments. But some animals, including eels and rays, seemed to produce a natural electrical shock. Was that the same thing again? Or was natural, God-made biological electricity a completely different force?

Galvani experimented on sheep and frogs, alive and dead. One day, he was working on a dissected frog near a machine for making static electricity. The frog was lying on a Franklin square—a foil-wrapped glass sheet that Benjamin Franklin had invented to function like a modified Leyden jar. Galvani mistakenly touched his scalpel to a nerve in the frog's leg. The leg contorted, twitching in rhythm to sparks emitted by the electricity machine. Galvani tried variations on the experiment, including some that involved affixing frog and sheep limbs by brass hooks to an iron rail. They jerked, but only when two different metals were in use.

Galvani concluded that he had discovered a new brand of electricity. He called it "animal electricity" and claimed that animals had an electrical fluid flowing from brain to nerves to muscles, the latter of which were de facto Leyden jars built into the body. In fact, what Galvani had made was an electrical current, a flow of electrical charges running through the metals in a circuit as the result of a chemical reaction.

While Galvani had legions of supporters, some fellow scientists were skeptical. Among them was Volta, a professor of experimental physics at Pavia University, a few kilometers outside Milan. He redid Galvani's experiments and soon realized that the trick was to have two different types of metals and some sort of moisture. It wasn't the animal's native electricity that caused the reaction, but rather the salty fluid in the animal's body that allowed electricity to flow, he said. He taunted Galvani, saying he didn't need his dead frogs, only some wet rags. But European scientists were divided, some fervently believing in animal electricity and others just as fervently rejecting the idea.

By 1800, after a decade of experimentation, and two years after Galvani had died, Volta made his breakthrough. He stacked pieces of zinc and copper on top of one another, separated by discs of cardboard soaked in salty water. Within this pile, a chemical reaction was taking place through the medium of salty water, stealing electrons from the zinc discs and depositing them onto the copper ones. As the electrons moved, they created electricity. It was dubbed "chemical" or "galvanic" electricity.

Volta had invented the battery. The batteries we use to power modern devices such as flashlights, cell phones, and even cars are the offspring of Volta's galvanic pile. In fact, the French word for battery is *pile*, a direct reference to Volta's stack of metals and briny cardboard. Batteries wear out over time because the chemical reaction becomes exhausted. If they are rechargeable, it means that the chemical reaction inside them can be reversed and the electricity can continue to be produced.

Volta lost no time writing up his experiments in the French language of the scientific establishment and sending the account to London to be published. A master of self-promotion, he feasted on the finding for the rest of his life, becoming a darling of Napoleon. He was one of the best-known, best-paid physicists in the world. Scientists all over Europe began making voltaic piles and running new experiments.

Still, confusion reigned over what, exactly, electricity was. Was galvanic, or chemical, the same as static and lightning? Were there other types, yet undiscovered? Ørsted, the newly minted academic, who in his 1799 doctoral dissertation had tried to argue that Kantian physics should include chemistry, made himself a portable voltaic pile and took it on the road. It was the latest thing, and everybody wanted to have a look at it, Jackson explained. So, in 1801, armed with the pile and a travel grant, Ørsted set out on a years-long international journey, using the novelty of the pile as a way of gaining entry to the best laboratories and drawing rooms in Europe, even scoring a meeting with the German writer Johann Wolfgang von Goethe. It was that trip that scotched his engagement to Sophie Probsthein.

It was a later version of Volta's pile that allowed Ørsted to plan his great experiment in 1820 at the University of Copenhagen, where he was a member of the faculty of medicine who liked to teach chemistry. He had been looking for nearly two decades for a link between electricity and magnetism, in keeping with his Kantian principles. In itself, this was both controversial and daring. Mainstream European science had roundly rejected the idea that there could be any connection. No less an eminence than the French physicist Charles-Augustin de Coulomb had declared that electricity and magnetism could not be related and there was no use looking for a link. Coulomb was the one who worked out the maths describing the law of electrostatic attraction and repulsion and after whom the standard international unit of electrical charge is named. ("There was precious little mathematics involved," Jackson remarked.) The French physicist and mathematician André-Marie Ampère, who gave his name to the unit of electrical current (often shortened to "amp"), ridiculed the very idea that electricity and magnetism could be linked.

Still, Ørsted persisted. In April 1820, he was freshening up a batch of lectures to give to a class of senior students and decided to devote a whole session to the elusive connection between electricity and magnetism. More than that, he would conduct an experiment in class to prove that they were connected. He hoped to do a trial run in private, but the events of the day overtook him and he didn't get around to it. On his way to the lecture, he hurriedly ditched the idea of doing the live proof, but once in class, overtaken with how splendidly things were going, opted again to run his experiment. He put wires to a voltaic battery and then to each other, creating an electrical current running in a circuit, and then moved it near a compass. The compass needle moved, but only feebly. The electrical current flowing through the wire was creating a magnetic field around the wire and the compass was reacting to that magnetic field. This had never been shown before. The students in the class were unaware that they were witnessing science history. Ørsted was crestfallen that the experiment had shown such unremarkable results.

Three months later, he tried again. He had decided that he needed a stronger battery, so he custom-made one for the task. It consisted of twenty voltaic batteries linked together to make their power add up. Each was a rectangular copper trough a foot high, a foot long, and two and a half inches wide, holding two copper strips. The strips were bent to hold a copper rod, which in turn held a zinc plate in the adjoining trough. Ørsted filled the troughs with enough water to nearly immerse the zinc plates and added slight amounts of both sulfuric acid and nitric acid. It was, in essence, Volta's pile turned on its side, making it more stable and capable of holding more fluid chemicals. He connected a wire to either end of the line of troughs—in modern terms to the positive and negative ends of the battery—and then connected the loose ends of the wires to each other. This was a closed electrical circuit, with the electricity running from the battery through the wire. The electricity was created by the chemical reaction between copper and zinc through the medium of the sulfuric and nitric acids. Simply put, electrons were flowing, making energy run through the wires. And there was so much of it that the wires themselves glowed with its heat.

Next, Ørsted suspended the conductive wire horizontally above the magnetized needle of a compass, in parallel with the needle. The closer the wire was to the needle, the farther west the needle pulled away from its usual north-facing position. If he put the conductive wire underneath the needle, the needle pulled east. No matter what type of conductive metal Ørsted tried for the wire, the compass needle moved. Even placing glass, metal, wood, water, resin, earthenware or stone—or combinations of these—between the wire and the needle did not prevent the needle from moving. This was indisputable proof: There was some sort of previously unrecognized physical connection between electricity and magnetism.

In all, Ørsted did sixty careful versions of the experiment. He was so concerned about how others would react to his findings that he conducted his work in front of eminent scientific witnesses who could vouch for his methodology. ("At this point does he know how important this experiment is?" I asked. "You bet!" said Jackson, nodding.) On July 21, 1820, Ørsted self-published his results in a sparely written four-page pamphlet—including the names and pedigrees of the witnesses—sent it by stagecoach to all the leading scientific lights and societies of Europe, and awaited the fallout.

The compass Ørsted used in the experiments is on display at the Danish Museum of Science and Technology in Elsinore, whose castle is famous for being the setting of Shakespeare's play about Hamlet, the melancholy Danish prince. The museum is in an unheated industrial barn reminiscent of an airplane hangar. ("They do not have so much there, and what they do have is not displayed well," Jackson had warned.) You get there by train and then bus, traveling north from Copenhagen through dark and forested northern European landscapes that bring to mind the gothic feel of "Little Red Riding Hood."

The compass itself is an elegant brass affair, covered by a glass dome, nestled on a carefully curved, highly polished dark wood base. You can see how it would have looked impressive to a class of students or an admiring group of Danish scientists. A replica of the elaborate battery Ørsted created for the experiments stands nearby on its own imposing wooden table. Two rows of ten copper galvanic troughs stand on its black-covered surface, dusted with the white detritus of chemical reactions. Affixed to the front end on either side are wooden spindles attached by wires to the ends of the pile. The wires attach to another set of spindles on the table and, finally, to each other, suspended overtop a compass. It's a huge and unwieldy apparatus, tucked into a drafty corner of the rather desultory museum.

Nearby, encased in a room made of glass, a display from Ørsted's laboratory and home gives a peek into his life. A box of glass and metal materials Michael Faraday gave to him. Rotating globes on high wooden stands. An elaborate candelabra he built and placed on his desk so he could work by candlelight. Photographs of his family. Shelves of his books, including two Bibles—one ancient, its brown leather creased with wear, and another, more stately in red and gold—along with his much-read copy of Sir Walter Scott's *The Lord of the Isles*. A copy of the poem Hans Christian Andersen composed to commemorate Ørsted's death.

Nestled in its own cubicle along one side of the display is a copy of the paper Ørsted published about his subversive findings. He wrote it in Latin, the formal language of science at that time, and entitled it *Experimenta circa effectum conflictus Electrici in Acum magneticam (Experiments on the Effect of the Electric Conflict on the Magnetic Needle)*. Carefully typeset, in rather large print for the day, the paper looks magisterial.

It caused an immediate splash, Jackson told me, rustling around on his bookshelf for a copy of one of Ørsted's works to give me. Triumphant, he pulled it off the shelf: a softbound, inch-thick volume, in both Danish and English, its glossy white cover adorned with the image of a middle-aged Ørsted, medals on his chest, hands folded over his stomach, looking prosperous and content. Called *Theory of Force*, it was his previously unpublished textbook in dynamical chemistry, unknown until a single proof from 1812 was discovered in an antiquarian bookstore in London in 1997. Jackson and Jelved tackled the translation a few years later and it came out in 2003. I had been planning to visit its publisher, the Royal Danish Academy of Sciences and Letters, at its neoclassical headquarters in downtown Copenhagen the next day to see if I could buy a copy. Jackson held it out, insistent: I must have this one.

The splash from Ørsted's 1820 experiment and paper was not just immediate. It was revolutionary. The unforeseen, unimagined, and inexplicable finding was that the magnetic force appeared to be circular. Above the magnet, the conducting wire forced the needle to the west. Below, to the east. That implied that the force was moving in a circle. Before Ørsted's experiment, the only forces that had been proven in science had worked in straight lines, explained Faraday's biographer, L. Pearce Williams. It "threatened to upset the whole structure of Newtonian science."

Within three months, Ampère had worked out a mathematical description, which still stands, of how electric currents give rise to magnetic fields, and then wrote to Faraday to ask him what he thought of it. ("Ampère was a very arrogant character," Jackson said.) Faraday couldn't read maths and demurred.

Jackson told me that when grilled by a rather chauvinistic friend the following February about why it was a Dane who made the discovery rather than the French with all their magnetic history, expertise, and equipment, Ampère wrote back blaming Coulomb. Coulomb had assured them there couldn't be a link and so, said Ampère, they didn't look. ("Never believe received wisdom!" Jackson advised with a dramatic shrug.)

Within a few months, Ørsted's paper had been translated and published from London to Paris to Geneva to Leipzig to Rome. Humphry Davy, the chemist who was president of the Royal Society in London, made sure Ørsted got the Copley Medal that year. Scientists all over the continent were reproducing Ørsted's experiments and some were conducting public demonstrations to convince the skeptical. By 1822, Ørsted had commenced what Jackson called a "triumphal procession" through Europe, meeting with scientists and discussing his grand finding. The same year, his insistence that chemistry be its own branch of science bore fruit. He was allowed to set up a chemistry lab that was untethered to the medical faculty and establish the position of full-time chemistry professor, a Danish first.

On a black-topped table in his office, Jackson had laid out his own apparatus to show me the experiment that gave Ørsted his place in history. He had a small compass in a clear plastic casing that could double as a tiny ruler, red string knotted at the top on the chance that you might need to hang it from your belt loop. Next to it was an unadorned black plastic case with two simple metal terminals inside, a plus and a minus, containing an AA battery. A black-plastic-coated wire came out the negative end and a red-plastic-coated wire out the positive. The whole contraption could easily fit in a trouser pocket. Jackson put the bare ends of each wire together—where they were stripped of their insulating colored plastic coatings—to make a circuit of electrical current and held them a few centimeters above the compass running in the same direction as the needle. The needle moved from due north to about 25 degrees northwest. No matter how often he connected the wires to make the current run, the needle still moved. The moving electrical charge was creating a magnetic field that the compass was responding to.

"That," he said, "is Ørsted."

No string of monumental copper troughs or glowing metal wires or diluted acids. Just a single everyday battery that today you can pick up at a corner store. And yet, as the science historian Gerald Holton put it, the finding "opened up physics itself to a succession of unifying theories and discoveries without which the modern state of our science would be unthinkable."

It was Faraday who figured out the next piece of the puzzle: Not only does a moving electrical charge make a magnetic field, but a moving magnet creates an electrical field. It is the basis for every electric power generator in use today. Jackson was set up to show me the nuts and bolts of that seminal scientific moment too. He held a foot-long plastic tube parallel to his body and inserted a strong magnet in its top end. The magnet swiftly fell out the bottom into his hand, just as you would expect. Then he replaced the plastic tube with one made of aluminum and repeated the experiment. This time, the magnet passed through the tube far more slowly than you would expect.

"And that," said Jackson, "is Faraday."

What was happening? Jackson explained: As the magnet moved, it created an electric current in the metal tube. That current created its own magnetic field, the equivalent to that of a magnet facing the opposite way. The opposite magnetic poles were resisting each other, and that's why it took the magnet longer to exit the tube.

But although Ørsted's finding was immediately accepted, his understanding of why it was happening was roundly rejected. Ørsted's Kantian interpretation, which he described as an electrical "conflict" between forces, made little sense to any of the eminent French and British researchers who tried to understand it. In fact, Ørsted seems to have had trouble describing what he meant and went back to the theme time and time again over the years, adding precious little more clarity. He spent three hours during a trip to Paris in 1823 trying to explain his ideas to Ampère and other French scientists. It was unsatisfactory. ("Ampère despised what he regarded as German speculative philosophy," Jackson commented.) Ørsted remarked in a letter home to his wife that the French didn't seem sympathetic to the idea of combining philosophy and science. In London, Faraday frankly admitted he didn't understand Ørsted's explanation, just as he had not understood Ampère's maths.

By the end of his life in 1851, Ørsted's abiding faith in Kantian natural philosophy had fallen out of step with the science of the day. No longer did scientists so roundly hew to the Romantic fashion of seeing God's design in nature. A more modernist and more empirical understanding was slowly emerging, paving the way for the findings of the late nineteenth and early twentieth centuries, including the discovery of atomic structure. Ørsted's magnum opus, *The Soul in Nature*, a floridly written philosophical dialogue that he tried to have published in English in 1848, was repellent to the few British scientists who read it. Charles Darwin, who in the next decade published his theories of evolution and natural selection, said he found it "dreadful." He spoke for all Britain. Ørsted, once at the forefront of scientific thought, was sidelined, most of his work spurned, if it was thought of at all.

CHAPTER 14

the bookbinder's apprentice

The Royal Institution, where Michael Faraday lived and conducted his experiments, is in Mayfair, the neighborhood of London that aristocrats favor when they are in town, much of it owned by the Duke of Westminster, one of the wealthiest people in the world. To get there, you might take the short stroll from Buckingham Palace up through the narcissus-dotted expanses of Green Park, one of the royal parks, before ambling past the Ritz Hotel on Piccadilly. From there, you head up Albemarle Street, named after the famously dissipated duke who owned the mansion that stood on that patch of London until he sold it to developers in the late seventeenth century to square his debts. Once on Albemarle itself you go past the art galleries featuring plush antique Persian rugs, past the flagship store of the American fashion designer Alexander Wang and the discreet shop of the designer Amanda Wakeley, who dresses the Duchess of Cambridge, and onward in front of the luxury jeweler Boodles until you arrive at number 21.

Michael Faraday first found his way there in the spring of 1812, just as Ørsted, over in Copenhagen, was preparing his ill-fated textbook on dynamical chemistry. Faraday was twenty and not at all part of the establishment that the area catered to. In fact, he was almost as far from an establishment figure as it was possible to be: a journeyman bookbinder with little formal education, destined by birth to be a tradesman. But he was fascinated by science and had taught himself some basics, primarily by reading books in the shop where he apprenticed as a binder. One of his most cherished was *Conversations on Chemistry*, by Jane Marcet. It was part of a series of illustrated introductory science books aimed at the popular audience, featuring conversations between the teacher, Mrs. B, and her two students, Emily and Caroline. This was not the scientific canon taught at universities.

The Royal Institution was a few years younger than Faraday was, set up in 1799 to put the "applied" into science for the sake of the expanding empire. There was agriculture to foster, mines and shipping to make safe with the latest scientific knowledge. As part of the impulse to democratize science and raise money, the institution put on lectures for the paying public. Faraday was there to hear one of them.

The lecturer was Humphry Davy, the Royal Institution's star attraction. Not only was he an engaging speaker, but his good looks had garnered him a substantial following among London's women, including Marcet, whose book on chemistry that Faraday so admired was based on Davy's talks. Davy's performances were so sought after that Albemarle was made into London's first one-way street in a bid to cope with the heavy traffic his appearances spawned. But the spring lectures of 1812 were to be his final appearances. The son of a Cornish woodcarver, he had determinedly and very successfully worked his way up the social ladder. He had been knighted that year and had come into money by marrying the exceedingly wealthy Edinburgh widow Jane Apreece a few days after he could bestow the title "Lady" on her. He was set to retire from the rigors of the public stage.

Faraday had acquired tickets to Davy's talks by chance, one of the most legendary bits of serendipity in the history of science. Tucked away behind a clock in the theater's gallery, he drank in the ideas and made careful notes. In the months following, he wrote up detailed accounts of Davy's lectures, adorned them with finely drawn illustrations, bound them, and, shortly before Christmas, got up the courage to send them to Davy. He had already served as Davy's copyist assistant for a few days after Davy had injured his eyes in a laboratory explosion. On Christmas Eve, Davy, clearly chuffed, wrote back an appreciative note. By March 1813, one of Davy's laboratory assistants had been sacked for being involved in a brawl and Faraday had taken his place at the Royal Institution, taking a pay cut from his journeyman's job to do so. He spent the following decades altering the course of science. Davy, who chemically isolated a string of elements, including sodium, in his chemistry lab, and who was not known for self-effacement, nevertheless once quipped that his biggest discovery was Faraday.

Few know Faraday's precise and brilliant trajectory through the scientific world better than Frank James, the Royal Institution's head of collections and professor of the history of science. James landed a job at the Royal Institution almost straight out of his PhD program at Imperial College London. And despite the fact that only half of one chapter of his doctoral thesis was on Faraday, he told me with a self-deprecating bow of the head shortly after we met, he became the editor of Faraday's 5,053 letters. It took twenty-five years and six door-stopping volumes to get through them all. Along the way James has produced many other books, essays, journal articles, and public lectures on Faraday and has pored over the bound notebooks Faraday made explaining the experiments he conducted throughout his working life. James has become so identified in the public mind with Faraday that a specially commissioned oil painting of him dressed in Victorian garb and sitting in Faraday's original magnetic laboratory now hangs in the Faraday museum in the basement of the Royal Institution.

I had written to James, asking to meet with him so he could help me understand how Faraday helped put together the concepts of magnetism and electricity in the wake of Ørsted's experiment. So, despite nursing a heavy cold that day, he was treating me to a morning latte at the institution's luxe café while we chatted. The building, which has been designated a historical site partly because of the work Faraday did within its walls, went through a budget-draining renovation in the first decade of this millennium, and the café now overlooks a glittering glass-and-steel elevator that pumps up and down through the open-concept heart of the building. Above us was a ring of shining offices. Below, enticingly, was the archive with Faraday's notebooks and the Faraday museum, whose refurbishment James oversaw. It includes the actual laboratory where Faraday did his magnetic experiments, which had been a servants' hall until Faraday took over on the quiet in the 1820s.

It was Ørsted's seminal paper, sent to Davy by stagecoach in 1820, that enticed Faraday into the world of electromagnetics, James explained. By then Faraday's genius had propelled him away from being a mere laboratory assistant and into doing his own experiments at the Royal Institution. Ørsted's paper had induced many others to write about electromagnetism and by 1821, everyone was confused. Faraday's friend Richard Phillips, editor of the journal *Annals of Philosophy*, commissioned Faraday to write the definitive review paper explaining to a hungry scientific community what this electromagnetic phenomenon really was.

So Faraday read everything he could find. It was a hodgepodge of contradictory information. He found Ørsted's talk of electrical conflicts simply confusing. He dived into Ampère's mathematical descriptions of electromagnetism, but he had never been trained in advanced maths and once described equations as "hieroglyphics." Ever the hands-on experimenter, he decided to repeat all the experiments described in the other journal articles, including those of Ørsted. Eventually he wrote up his findings into a series of articles for

Phillips, signing his name for public consumption humbly, and mysteriously, as "M." Here was the first understandable explanation of electromagnetism that the world had seen. The articles were so popular that the public implored their author to unmask himself. Faraday did so and tasted fame.

But as Faraday repeated the experiments of others, he had thought up some of his own. What caught his attention was precisely what had made Ørsted's compass needle move when it was near the current of electricity. The traditional thinking, which Ampère supported, was that the needle and the wire were being attracted and repelled by each other, that the power between them leapt across empty space and distance. Faraday began to wonder whether there was instead a circle of force in the space around the wire—a physical thing—that was affecting the compass needle. That could be one way to explain the peculiar circular effect Ørsted had observed.

Being Faraday, he devised an elegant experiment to test the theory. On September 3, 1821, he took a basin, a glob of wax, an iron magnet, and a quantity of quicksilver, or mercury, which is a conductor of electricity. He fixed the magnet, north end up, to the bottom of the basin with the wax and then filled the basin partway with mercury. Then he hung a piece of wire from an insulated stand so it could swing freely in the mercury around the magnet. Finally, he created a closed electrical loop by connecting a battery to the wire on one end and the mercury on the other. The wire moved clockwise around the magnet. The electrical current running through the wire created a magnetic field. That magnetic field interacted with the field surrounding the magnet, causing the wire to rotate around the fixed magnet. Then he reversed things. He loosed the magnet and fixed the wire. The magnet could float in the mercury, attached by a tether to the bottom. The wire was immobilized in the center of the basin of mercury. The magnet revolved around the wire as soon as Faraday made the current run.

This was the first electrical motor: the creation of mechanical energy from the power produced by an electrical current and a magnet. Faraday called it an electric magnetic rotation apparatus. He seems to have had no precise concept of what such a machine might be made to do—he could not have foreseen the mechanization of the world that we now experience—but he knew it was important. For one thing, it reinforced his odd idea that magnetic forces might curve around the magnet, filling space. His summation, entered into his laboratory notebooks describing the experiments, reads: "Very satisfactory, but make a more sensible apparatus."

It would be a decade before he had the time to turn his attention back to the puzzle of electromagnetism.

CHAPTER 15

magnets making currents

Sipping a latte in the building that is renowned because of the work Faraday did within its walls, it's hard for me to take in just how much of an outsider he was when he began his career. Today, Faraday permeates the place. The elegant instruments he made by hand, the controlled-access archive of his notebooks, and the enduring lore of his story are mainly what draw people here. A statue of him stands rather grandly at the base of the building's curved staircase, his austere form draped in academic gowns, grasping a replica of the ring coil, his most famous apparatus. A bust of Faraday by the nineteenth-century sculptor Matthew Noble is in another part of the building. Margaret Thatcher so idolized Faraday—he wasn't born with a silver spoon in his mouth, like her, and studied chemistry, like her, and made good, like her—that she borrowed the bust in 1982 after she became prime minister and made it the first thing visitors saw when they entered 10 Downing Street. One hundred and seventy years after he set foot in the Royal Institution on charity tickets, Faraday had been pressed into service as the prime symbol of British can-do.

The change in his status was remarkable. It wasn't only that

Faraday had been born into trade rather than marked for a scholarly life. It was also his religion. Like his journeyman blacksmith father, Faraday was a Sandemanian. It was a straightlaced Protestant sect, born by way of dour Scottish Presbyterianism. Its adherents met each Sunday to ritually wash one another's feet and feast together in memory of Christ's life and sacrifice. They preferred one another's company, cautious about too much socializing with others and often marrying within the faith. Ever mindful that the ultimate reward was a place in the kingdom of heaven, they eschewed the riches of the world, preferring plain living and the practice of sharing what they had with the poor. They knew that they could count on salvation, and it fostered serenity, even joy. "A peculiar aura of good nature" surrounded the Sandemanian group, wrote Williams, Faraday's biographer.

In Faraday's time, not being Anglican, the official Protestant denomination of England, was a severe handicap, James explained, episodically wiping coffee from his full, walrus-style mustache. Professors at the Universities of Oxford and Cambridge, had to be Anglican by definition. Naval and military officers, some of whom were also natural philosophers, were also required to be Anglican, apart from a few exceptions for Catholic Ireland. Not only that, but being a scientist in itself was somewhat countercultural. By James's calculation, only about a hundred people in Britain were paid to do science in 1812, the year Faraday went to the Royal Institution to hear Davy's lectures. Many who devoted their lives to it were independently wealthy members of the landed and titled classes. Some, like Davy, wanted to join that elevated echelon. Not Faraday. Twice Faraday was offered the esteemed position of head of the Royal Society and twice he declined. His concession to glory was to accept a "grace and favour" home outside London late in life from Queen Victoria's consort, Prince Albert, where he lived at the Crown's expense in his final years.

Yet while Faraday's religious faith made him an outsider, it was also one of the reasons he succeeded. He saw things differently from others. And he had a different reason to pursue his experiments. Faraday did science in order to understand the world he believed God had created. In turn, his belief in God informed his understanding of science. It was a slightly different philosophy from that of Ørsted, who believed in an almost pantheistic philosophy, that every facet of nature showcased God's greatness. Instead, Faraday perceived the presence of intricate, shadowy, and even tricky godly laws that were responsible for everything around him. Discovering them would take ingenuity, and a lifetime.

That same religious faith prevented Faraday from believing in atoms. That idea ran against his understanding of how God had created the world. What today we see as a combination of atoms, or a molecule, Faraday saw as a solid piece of material that could be divided into yet smaller pieces of itself. He had no concept of the airy interior of an atom surrounded by electrons that Bohr came up with in the following century. For similar reasons, he loathed the term "scientist," James told me. It comes from the Latin *scientia*, which means "knowledge," and Faraday believed the term stripped God out of why the world was there in the first place. Faraday preferred to be called a "natural philosopher" or a "man of science," James said.

People were never quite sure how to react to Faraday, James told me, getting up from the café table to show me to the museum in the basement. They didn't know where his ideas came from. Perhaps because he had such an atypical background, he had the knack of being able to envision how natural forces might interact with an apparatus and then think up experiments to test those visions, regardless of what traditional scientific theory dictated. All of it set him apart. But he was correct so often, James said, that they couldn't dismiss him.

At the entryway to the museum was a display containing the homemade piece of equipment that Faraday is most famous for, adorned with pumpkin-colored explanatory signs. Dated August 1831, it was

Faraday's most inspired attempt until then to make electricity from magnetism. Faraday had spent most of the 1820s mired in a project Davy had pressed on him, unable to work on other experiments. Davy was chairman of the Board of Longitude and, despite Harrison's H4 clock of 1759 that had technically resolved the longitude problem, the British Admiralty was still trying to figure out a cheap and easy way for sailors to figure out where they were. The board had placed its money on reading the heavens, and Davy had decided that Faraday could do the important job of making better optical glass for sailors' telescopes. It was miserable work. He had to have a glass furnace installed in his laboratory. Later in life, Faraday spent a great deal of time trying to prove that secular variation of the magnetic field was linked to fluxes of oxygen in the atmosphere as it warmed and cooled over the course of days and seasons. (He was wrong.) But at that time, he was not much interested in terrestrial magnetism or longitude. He abandoned the project as soon as he decently could after Davy died in 1829, and resumed the electromagnetic experiments that were to consume the following decade of his life.

Again, Ørsted's pioneering ideas were an influence. During his European trip of 1801, Ørsted had seen some of the astonishing images created by the German physicist and musician Ernst Chladni. Chladni had run a violin bow across the edge of a metal or glass plate on which he had scattered sand. The waves from the vibrations made geometrical patterns. It was like being able to see sound. By 1806, Ørsted had begun to think about whether sound and electricity might make similar patterns and did his own experiments using fine moss seeds on plates. Faraday read about his work, and in early 1831, added light into the mix. Could all three—sound, electricity, and light—be made up of vibrations? He set up a six-month series of acoustic experiments to test the idea. His astonishing conclusion was that the sound vibrations existed in the air. It put paid to the idea that the air was empty. From there, it was a short step for Faraday to begin thinking about electricity as waves. By late summer, he started experimenting.

A t that point in the history of electromagnetism, anyone could make a magnet from electricity. All it took was placing a piece of iron inside a coil that thrummed with electrical current and the iron became a magnet that held its charge. Faraday wanted to do the reverse: make electricity from a magnet. As a first step, he commissioned a wrought-iron ring, seven-eighths of an inch thick and six inches from outer edge to outer edge. In his mind's eye, he divided it in half. Around one half, he wound three twenty-four-foot-long pieces of copper wire—the same type then used in making bonnets—as tightly together as he could. The more turns the wire took, the greater the magnetic effect of the electric current would be. He insulated each turn of coil from the next with string, as insulated wire was not available in 1831. He kept going, making another layer of wire coils on top of the first, insulating each layer with cotton calico dress cloth. Then he performed the same process on the other half of the ring, leaving spaces on the ring between the two coiled sides. James told me that he has estimated it would have taken about ten days to make the apparatus.

Today, you can see from the display that the calico is discolored and a bit tatty. Pieces of the string have sprung loose. But the experiment Faraday conducted with this rather homely device is one key to the vast electrical infrastructure that carries electrical currents around the world. It was the first transformer, capable of making fast-moving electrons pushing hard at high voltages slow down enough to be useful in everyday, low-voltage applications. Today, transformers in power stations allow the whoosh of electricity generated by water, sun, wind, nuclear reaction, and coal to flow at lower speeds into your lamps and computers and other devices run by electricity.

On the day he did his experiment, Faraday hooked up one side of the wire-coiled iron ring to a battery with a switch. He hooked up the second side to a galvanometer, a device to measure electrical current. He flipped the switch. Electrical current from the battery flowed through the copper wire on one side of the ring, producing a magnetic field. The wire on the other side of the ring briefly made the galvanometer flick with a small pulse of current. Then it went back to its neutral position despite the strong current that continued to flow on the other side. When Faraday switched the battery off, the other side again showed a small, brief burst of electricity, but in the opposite direction on the galvanometer. Faraday's conclusion was that electricity was created not when the magnetic flow began but when it changed. The electricity created on that other side was also lower-voltage than the original battery current, the basis of the device's usefulness as a transformer. The device has gone down in history being called an induction ring, meaning that it had induced an electrical current, however intermittent.

But why did a change in magnetic flow make electricity? And could electricity be made using a magnet alone, without the current from a battery? Three months later, Faraday found some answers. It was a simple experiment, the same one Andrew D. Jackson replicated for me in his office at the Niels Bohr Institute in Copenhagen. Faraday took a hollow iron tube and wound its outside with copper wire insulated with cotton. Then he hooked the tube up to a galvanometer and slid the magnet through the tube. The galvanometer recorded a current. When he slid it the opposite way, the galvanometer recorded a current in the opposite direction. Just a twitch. As the magnet moved it created an electrical current, pushing electrons in the wire surrounding the tube. That electric current in turn created its own magnetic field. Faraday had created a generator of electricity without using a separate source of electricity.

James and I had arrived in a hallway in the basement. He stopped at a display of the pile Volta gave to Faraday in Milan in June 1814, when Volta was nearly seventy. It is one of the institution's treasures. Among the earliest batteries ever made, it looks unassuming, standing perhaps a foot high on a highly polished wooden base. Davy wanted the pile to be bigger and stronger and more physically stable, so he came up with the innovation of turning it on its side, James said, chuckling. That led to the troughs Ørsted used in his grand experiment of 1820.

Farther along, past the painting of James in period costume, was Faraday's magnetic laboratory, arranged to show what it had looked like in the 1850s toward the end of his working life. The laboratory was preserved, not because Faraday was such a legend in his own time but by accident. After he died in 1867, nobody bothered to clear out the materials inside. But by 1931, the hundred-year anniversary of Faraday's discovery of the transformer and generator, he had become a symbol of the British role in creating modern technology. People began unpacking his old laboratory, now a basement storeroom, and found his instruments, chemicals, vials, even the dumbwaiter where servants had long ago loaded things to be carted to higher floors. Faraday used the dumbwaiter to store experiments and its doors still bore the red wax seals he had placed there to signify that he had experiments locked inside.

Faraday's experiments and findings in magnetism and electricity ranged much further after these early successes. One of his contributions was to establish unequivocally that all forms of electricity were identical, no matter how they were produced, James explained. At that time, scientists thought of electricity as coming from five different sources: static (known as "common"), voltaic, animal, lightning, and thermal. Did each type produce the same results in experiments? Faraday painstakingly went through what was known about the qualities, or "identity," of each and ran experiments to test each type, filling in a chart as he proceeded until he proved they were the same thing.

Each test reinforced his idea that lines of force filled the air. It's the same idea grade school students see during the experiment with iron filings arranging themselves in lines across a paper overtop a bar magnet. On April 3, 1846, Faraday introduced the idea in an extemporaneous public talk, one of his famous "Friday evening discourses" at the Royal Institution. He had arranged for another speaker to give the lecture, but that person had bolted due to nerves. Faraday hurriedly took the stage in his place but found himself with twenty minutes to spare once he had finished his topic. So, for the first time in his life, he went off book. In retrospect, it was an extraordinary moment in science. Faraday summoned up a vision of the world filled with lines of electric and magnetic force, and perhaps even other forces such as gravity. They had physicality. They formed matter. He was describing the electromagnetic field, other fields, and some of the concepts that would eventually underpin quantum field theory. But while he could describe what he had found in precise, sometimes bespoke, language, he lacked the ability to describe his findings in the universal language of physics: mathematics.

CHAPTER 16

the lines that fill the air

The unconventional Scottish physicist James Clerk Maxwell did understand mathematics. He read Faraday's papers on electricity and magnetism and then, by pulling together all that was known about them, wrote four new equations in a paper published in 1861. For the first time, they described electromagnetism.

This was a far more difficult task than simply translating Faraday's words into maths. Maxwell had to work out that not only does an electric current produce a magnetic field but so does an electric charge that is only displaced, not flowing continuously. Contained within this idea was the maths to prove that space is filled with electromagnetic waves independent of any electric currents. Electricity, therefore, was a manifestation, even a subset, of the electromagnetic force. This was something the electricians of the previous century could never have imagined. When Maxwell read his equations carefully, they also told him that electricity, magnetism, and light were aspects of each other. They all behaved as waves traveling at the speed of light across space, a speed he had calculated a few years later. This was the collection of invisible lines of force that Faraday had dreamt of and partly glimpsed. Now it was enshrined in mathematics, available for all physicists to play with.

Maxwell's equations predicted that electromagnetic waves can be literally any length. As the physicist Neil Turok explains, the waves are just "stretched-out or shrunken-down versions of one another." Most are invisible to humans, just as most frequencies of sound waves are inaudible, including the high frequencies of ultrasound. The electromagnetic waves we can see are far smaller than a millionth of a meter in length and they give us color. The very longest we can see are red; the shortest, violet. But there are even smaller electromagnetic waves, such as the dangerous gamma rays produced in the Large Hadron Collider, whose activities were lighting up the façade of the Niels Bohr Institute in Copenhagen when I was there, and X-rays. Extremely long waves, known as ultra-low-frequency waves, can penetrate the Earth and are used to communicate in mines. Other types include microwaves, which work in the appliance of that name but also make radar function. There is also the big group longer than microwaves called radio waves, which are put to work in cell phones, radios, and televisions. But as different as they may seem, all these waves can be described mathematically in exactly the same terms. That finding laid the groundwork for the electrical infrastructure that supports virtually all the energy and information we use in our modern world.

Ultimately, Maxwell's equations led to the elegant maths describing the standard model of physics developed in the early 1970s. Today, if the standard model equation proves that something can be true, then it is, no matter how counterintuitive it might seem. That's how an electron can be a particle and a vibration in a field at the same time, or how the Higgs boson was imagined before it was found. The revolution in scientific thinking this represents is vast. At the beginning of electromagnetic research, the Bible was the only truth. Natural philosophers, like Galileo, had to flout authority to make observations about nature that went against what the Bible said. Data points and logical conclusions were dangerous. Later, observational results were all that mattered; the highest scientific endeavor was to explain the world through the evidence of your own eyes. Today, the standard model equation is king, with its alarming precision and its preposterous abstractions. Observation, while not passé, is not everything.

Here's an example: Maxwell's equations theoretically connected space and time, a profoundly improbable fact. That led directly to Albert Einstein's special theory of relativity, which states that time and space are not fixed. Time does not march on, unheeding, as the poets might say. Neither are time and space separate from each other. Instead, they are a continuum, capable of adjusting themselves in order to make sure the speed of light—that is, Faraday's tiny electromagnetic wave—is fixed.

Einstein published his special theory of relativity in 1905 in the journal *Annalen der Physik* while working in a patent office in Bern, Switzerland. It was one of four remarkable papers he published that same year, known as his *annus mirabilis*, or miraculous year. His work that year changed the way physicists saw time, space, mass, and energy. That same year, a few hundred kilometers to the southwest, Bernard Brunhes made his trek by horse to Pont Farin in the Cantal of France to the brand-new roadcut, where he hacked away at the seam of ancient terracotta that showed that the Earth's poles had once been on opposite sides of the planet.

I had made a special request that James show me the entry in Faraday's journal from the fateful day in 1831 when he wrote up his experiment with the induction ring. James pressed a code pad that allowed him into the locked archives, home not only to Faraday's notes but also to those of Davy and other scientists who have worked at the Royal Institution over its centuries. Other researchers were ensconced there, along with the keeper of the collection. We were in the basement, just a few meters from where Faraday had worked in his magnetic laboratory.

Box after box of scientific treasure sat on the metal shelves, tidily labeled. It was the emotional and perhaps spiritual core of the institution. Faraday's notebooks were in sturdy, flattish brown cardboard boxes with removable lids. James, with an ease born of long practice, consulted the shelves for a few moments and then took one down and removed its lid. Inside was an elegant oblong brown leather notebook inscribed with gold lettering, evidence of Faraday's early passion for bookbinding. James opened the volume to August 29, 1831, and held it out with a slight flourish.

There, in sepia ink, in a sedate and beautiful hand, was Faraday's description of the experiment that set him on the road to discovering that a magnet plus movement could make electricity, opening up the largely invisible world of electromagnetic fields snaking through the universe. Lines of script evenly spaced on a single page, written with scarcely a correction, a tidy diagram of the induction ring midway down the right-hand side.

It seemed to me that I could see Faraday writing the lines, sitting upright and proper in his laboratory. I could imagine him puzzling over the mysteries that were slightly out of sight. And from him I could trace a path to Maxwell working out his four famous equations at his Scottish estate south of Glasgow, and from there, to Einstein in that patent office in Bern reimagining the nature of space and time, and then over to Brunhes in his slender Renaissance tower in Clermont-Ferrand, realizing that the Earth's magnetic field was far more mercurial than anyone had imagined.

About a hundred years later, their work, plus that of hundreds of other scientists, would unveil an electromagnetic reality that was breathtakingly more inconstant. Despite all that we now know thanks to advances in quantum physics, particle physics, geophysics, mathematics, computers, and satellite technology, we cannot predict how the Earth's magnetic field will behave. That ability remains resolutely out of reach. But we have a few pieces of evidence. We know that the field is decaying more rapidly than many scientists had predicted. It is more unstable in the south. And if the poles are again preparing to switch places, the infrastructure that carries electrical current to our doorsteps is in danger of being damaged beyond repair.

PART III

core

And what rough beast, its hour come round at last,
Slouches towards Bethlehem to be born?

—William Butler Yeats, "The Second Coming," 1919

CHAPTER 17

the contorting gyre

Despite the sweltering heat of the summer evening, the theater at the convention center in the French city of Nantes was filling up with people. They were there to hear about the journey to the center of the Earth that is, even now, in progress. The topic was a natural for citizens of Nantes. It fed into the lore of the city, once home to the writer Jules Verne. His science-fiction novel about an irascible geologist's pilgrimage into the Earth's core was published in 1864. Verne's fictional geologist climbed up an extinct Icelandic volcano from which he descended straight into the Earth's bowels, stumbling on an underground ocean populated with ancient monsters before finally making his way to the surface again. But today's voyage to the underworld is not literary. It is scientific. It is fraught with implications for life and human civilization.

Philippe Cardin was on the stage in Nantes. A physicist at the University of Grenoble in southern France, he was one of the exclusive and tight-knit group of the world's scientists who study the machinations of the Earth's deep interior. Most of them, which is to say, about two hundred, were gathered in Nantes for the conference they hold somewhere in the world every two years to share their newest findings. This was the second day of a week of revelations. Cardin had been tapped to give the week's only public lecture.

He was a gifted storyteller. The audience sat rapt as he romped his way through the history of how scientists have come to know about the enigma in the heart of the planet. He wanted to make it clear that for him, it's not just about what's there now. The art of reading the entrails of the Earth means reading its birth pangs, its evolution, and its future. It is to acknowledge that the Earth is alive, that it has undergone immense change, and that it must continue to change. In Cardin's field, a longed-for goal is to be able to read what will happen next. It is elusive.

He made a confession to his audience that surprised me. He put today's scientific explorers of the Earth's core into the same category as the literary explorers of old. Not just Verne, whose tales of extraordinary voyages still make him one of the most beloved novelists in the world. But also Dante Alighieri, who, in the fourteenth century, helped establish the idea within the collective human imagination that the descent into the underworld was a descent into hell. Dante banished the arch-devil Lucifer to the ice-fast lowest level of the deepest circle of suffering—mute, immobilized, abandoned. Other artists have depicted going into the underworld as the opposite: the physical and psychological search for a sacred place, where there is protection, warmth, and wonder. Scientists too, Cardin told his audience, were drawn into the earthly abyss by the allure of imagining the unknown. The quest captures them and they cannot let it go.

The culture of the scientific world doesn't lend itself easily to this sort of disclosure. Scientists rarely acknowledge the psychological pull their subject has on them, or the torrent of imagination their work entails. I caught up with Cardin the day after his lecture, fascinated by his willingness to name the emotional seduction of science. As we walked along the Loire River that runs through Nantes, I told him that Jacques Kornprobst had taken me to Pont Farin to show me

Bernard Brunhes's field-twisted terracotta. At once, he became voluble. He knew Kornprobst as a valued senior statesman of French geophysics. And he was unusually well informed about Brunhes, despite not having mentioned him in his talk the night before. ("Ah, yes," he said. "Perhaps I should have talked about Brunhes!") A few years earlier, Kornprobst had invited Cardin to give a public lecture at Vulcania, the European park of volcanism near Clermont-Ferrand, to honor the centenary of Brunhes's discovery. It had been Kornprobst's article on that meeting that drew me into his sphere. To prepare for the Vulcania lecture, Cardin had reread Brunhes's work. Cardin was still marveling at what Brunhes had been able to envision back in 1905. It was like looking at a river—here he pointed to the Loire—and imagining the ocean without ever having seen it.

Seeing the invisible has been a hallmark of geophysicists over time: How did the Earth come to be? How was that rock made? What is in the center of the planet? Why? And catching a scientific, rather than literary, glimpse of the architecture within the center of the planet has had its own peculiar narrative arc. It started with William Gilbert, who claimed in 1600 that he was the first to take a scientific look into the planet's innermost reaches. Gilbert concluded from his experiments with terrellae, or lodestone models of the Earth, that our planet is a giant magnet with a magnetic soul that shuttles power from deep inside it all the way across its surface. It was an inspired guess then, and happened to be correct. Incorrect was his idea that the Earth's great magnet was responsible for its ability to spin. In fact, the Earth's need to shed heat from its inner domains makes the electrical currents that produce the magnetic field. Its spin helps to organize the field into the simple two-pole structure that allows us to use it for navigation. Later that century, Edmond Halley pushed Gilbert's ideas further, positing that the core might be liquid and that changes within it are responsible for the magnetic soul's constant change. That's become the prevailing scientific consensus.

From the center, the magnetic field spreads in unending loops from pole to pole, stretching tens of thousands of kilometers beyond Earth's surface, forming what's called the magnetosphere, which surrounds our planet and interacts with other magnetic fields in the universe, including the sun's. The Earth's vast electromagnetic field creates a protective shield around our planet, fending off solar wind and cosmic rays. On the side facing the sun, it is squashed flat by the violence of high-energy particles emanating from our mother star. The field pinches in at either pole where the loops run through the Earth's center and then streams out the side away from the sun in protean tentacles reminiscent of a galactic squid. This odd creation, which developed when the Earth was perhaps as much as a billion years old, guards our planet from the ravages of radiation. It may even be the reason life exists on Earth.

But how is that electromagnetic power generated within our planet's core? Like everything else in our universe, its origin is violent. One critical component is the unpaired spinning electron.

About 4.6 billion years ago, our solar system was just a cloud of dust and gas nestled within the universe. Then something happened—perhaps it was a nearby star exploding—to make the cloud collapse in on itself. The result was a flattened disc of gases and bits of debris, some of which collected in the middle, eventually picking up enough density to start nuclear reactions. That was the birth of our sun.

But while the infant sun consumed most of the matter in the system, there was still a great deal of material rotating around it. As bits bumped into one another, they clumped together to form hundreds of thousands of tiny planets, called planetesimals. These were the forebears of the planets in our solar system: Earth and its siblings. Some were icy, some rocky. Near the sun it was too hot for all but the four rocky protoplanets that were to become Mercury, Venus, Earth, and Mars, and so they began to orbit closest to the sun. Farther out were the gassy formations that became Jupiter and Saturn, and farthest of all were the two dominated by ice, Uranus and Neptune (and perhaps the mysterious potential Planet Nine that scientists recently caught a theoretical hint of in the Kuiper Belt, or yet another in the theoretical Oort cloud).

Still, the planets were accreting. As they got bigger, their gravitational pull got stronger, so they attracted even more material toward themselves. The material crashed violently into the growing planets, generating heat from the force of the collisions. That heat was locked in the planets' cores like a giant furnace containing the savagery of their birth.

Among the materials forming Earth and the other rocky planets was lots of iron and a little nickel. But because iron and nickel were among the heaviest elements produced in this embryonic solar system, they sank to the center of the planet while lighter material accumulated on the surface. The Earth's center was so hot that it kept the metals liquid—too hot to carry a magnetic charge because they were above the Curie point that so fascinated Brunhes. By the immutable rules of physics, heat captured within the molten metal core needed to find a way to escape, so the Earth began shedding heat from the inside out, just as a pot of water throws off heat in the bubbles it produces across its surface when it boils. As the core succeeded in discarding heat, its innermost portion solidified, leaving the outer core liquid.

That liquid metal of the outer core was still churning, helping to shunt heat away from the inner core while it rotated around its axis. That same axial rotation caused the inner-Earth liquids to dance to what's known as the Coriolis effect. Named after a nineteenth-century French mathematician and engineer, the Coriolis force governs the spin of large bodies of liquids on the planet, including the ocean, the atmosphere, and the molten metal in the outer core. That's why ocean gyres and hurricanes in the northern hemisphere run counterclockwise and those in the south run clockwise. They are curving as they move across the Earth's surface. In the outer core, the

Coriolis effect creates discrete north-south-running columns of liquid metal spinning on the edges of the solid inner core, another splendid mechanism for getting rid of excess heat. In addition, the atomic construction of the iron and nickel in the core makes them excellent conductors of electrical charge. Each iron atom has four unpaired electrons spinning in its outermost filled orbitals and each nickel atom has two.

What that added up to, in this early Earth, were the magic twin elements needed to produce a dynamo—a generator of electricity that can sustain itself for billions of years: heat energy carried by a liquid, plus an electrical conductor. The dynamo produced electrical currents that flowed through the molten metal. Electrical currents, as Ørsted later showed with his copper troughs in Copenhagen, produce a magnetic field. The field that the dynamo began to produce billions of years ago cocoons us still. It is an artifact of our planet's birth.

If either ingredient had been absent, the Earth would not have a magnetic field. For example, if the core were not iron and nickel or other superb conductors of electricity. If the core had cooled down enough so that it did not have to transport heat energy. (As it happens, the silicate mantle that Kornprobst loved is like a blanket, helping to keep enough heat in so that the core sheds heat relatively slowly, keeping the dynamo alive.) Today, the innermost solid part of the core is about two-thirds the size of the moon and is growing slowly. It is still at a high temperature—somewhere in the range of 5,000 degrees Celsius—but because it is under tremendous pressure, its "freezing point," or point of solidification, is also high. The outermost core, which is under less pressure, is like a runny liquid—at about 4,200 degrees Celsius—sustaining the dynamo. In billions of years, when the outer core finally cools enough to solidify, the Earth's magnetic field will die. It will be like Mars. Mars had an internal magnetic field until, a leading theory suggests, its core cooled enough that the dynamo, and therefore the magnetic field, sputtered out. It still has a weak magnetic field in the rocks of its crust. But without the strong internal field, solar wind ripped away Mars's atmosphere, which may be one reason there's no life on that planet. Likewise, our moon once had a dynamo and now has a weak magnetic field that resides in its crust. Tantalizingly, rock records on both the moon and Mars retain a chronicle of the history of their internal magnetic fields, if only we could get at them and read them the way Brunhes once read ours.

Other planets still have magnetic fields. Mercury, the tiniest rocky planet. The gas giants, Jupiter and Saturn. Both ice giants, Uranus and Neptune. So does the sun. The sun's dynamo is driven by the extreme heat given off by nuclear fusion going on within its interior. The heat is so intense that it knocks electrons out of their orbitals and they move in electrically conducting plasmic waves, creating a magnetic field. The sun's magnetic field and ours are constantly interacting. When the sun has episodes of high magnetic activity, our field lets more solar wind into the upper reaches of our atmosphere and we often get brighter auroras that show up at lower latitudes. When our field weakens, our magnetosphere compresses, again, allowing the voracious solar wind to penetrate closer to the Earth's surface.

All these celestial bodies rotate. A rotating body reinforces a two-pole, or dipole, magnet, roughly akin to a bar magnet running through the center of the planet or the sun, aligned with the angle of rotation. The dipole is the default position. Sometimes the direction of the field needs to change. But because the field must run from north to south, when a field changes direction, it means the poles have to switch places too.

In the case of the sun, the poles change places every eleven years as the magnetic field is annihilated and reborn. It's a highly volatile system. The sun feeds on the change that a reversal causes in order to keep going. Scientists know this because the sun has no rocky crust. It is naked. They can see inside it, tracking reversals in the direction of the field and the shifting position of the poles. Solar pole flips coincide with more sunspots, which are small patches of the sun's surface that are cooler than the rest and therefore look darker. We have precise records of their patterns going back five hundred years, including a whole month's worth drawn by Galileo in 1612.

The Earth's rotation gives it a bias toward a two-pole magnetic field too, which we implicitly refer to every time we talk about our North and South Poles. As with the sun, our magnetic poles have switched places many times through our planet's life. But unlike the eleven-year cycle of the sun, our planet's poles take eons between reversals: roughly every 300,000 years in recent eras. The last time they flipped was 780,000 years ago, leading some physicists to ask whether the next reversal is overdue. Is the core plotting to shift the direction of the Earth's magnetic field sooner rather than later?

Recent satellite findings have given them new cause to wonder. A more complex picture of the field within the core has emerged. Yes, the Earth has a dipole. Yes, the dipole is dominant. But the field has more submissive and more complex structures too. The triplet of satellites in the atmosphere now monitoring the Earth's magnetic field—the Swarm satellites launched in 2013 by the European Space Agency—are tracking an epic battle going on inside the Earth's core. Possibly fed by a gyre in the outer core, non-dipole magnetic fields are struggling to topple the governing dipole. Their power is reaching into the dipole, striving for insurrection. It is like the ancient battle of the titans, taking place in the underworld. Already, a chunk of the dipole field in the southern Atlantic Ocean, ranging roughly from Africa to South America, below the equator, has succumbed. There, the field is running in the opposite direction from the way it is supposed to be running. The protective capacity of the magnetic field overtop that part of the Earth has decayed dramatically enough that communication satellites circling overhead shut off as solar radiation attacks them. The field is not strong enough to repel the sun's danger-ous radiation in the atmosphere—it remains strong at the surface—in that part of the world. The field has waned in unexpected ways.

A question for Philippe Cardin and the other experts here at the conference in Nantes is whether that waning will continue and spread or whether the dipole will fend off the interlopers. The Earth's dipole is already weakening, although slowly. When it gets weak enough—whether it's in this go-around or later—and when the rearguard non-dipole magnetic factions in the core challenge the dipole's dominance strongly enough, the two poles will falter, as they have hundreds of times before. The poles will be thrust from their current positions. Other poles from the non-dipole fields will gain strength. The Earth will have perhaps four or eight magnetic poles during that time of transition. Our protective shield will wither to only one-tenth of its normal strength during the time the poles are traveling. That process could take thousands of years—meaning, here on Earth, we could be exposed to more radiation for those same thousands of years.

At some point in the future, the two poles will find themselves on opposite sides of the Earth from where they are now. The default dipole field, reinforced by the rotation of the Earth and the Coriolis force, will regrow itself, snapping back into place. But the direction of the field will have changed. What we think of as our current north pole will be in the south. South will be north.

We don't know when the reversal will happen or how long it will take to complete. We don't know exactly what will happen to life on Earth during the process. But we are gathering hints about how this unpredictable system in the planet's core works, and glimpses of why it does what it does.

CHAPTER 18

shocks inside the earth

The monsoon rains had been falling for two days when the earthquake hit, and the ground of the Shillong Plateau in northeastern India was saturated. By 5:15 on the afternoon of June 12, 1897, when the earth began shaking, the land was so wet that much of it melted away beneath people's feet, a process geologists refer to as "extensive liquefaction." Land slid. Bridges sank. Sand boils and mud volcanoes erupted. Every single building in an area about the size of Louisiana was reduced to rubble. A cleft that ran for miles cracked open in the subterranean crustal plate. More than 1,500 people died. Known as the Assam earthquake, it has been estimated at a magnitude of 8.7 and is considered one of the largest in modern history.

A dozen primitive seismographs in Europe captured the movement of the Earth as the crust ruptured, tracking the waves of shocks and aftershocks as they flowed from one side of the planet to the other, through the center. Seismology was experimental at that time and geologists were only beginning to be able to read the story the graph lines could tell. But the Irish geologist Richard Dixon Oldham happened to be in India working with the Geological Survey of India just then. All corners of society dissuaded him from investigating the huge earthquake. They had different priorities, being fixated not on death and destruction but on the Diamond Jubilee celebrations of Queen Victoria, just eleven days away. Nevertheless, Oldham went to the site, looked at the seismographic records, and produced a carefully written report. But he kept thinking about those waves, and by 1906, the year of Brunhes's paper on the terracotta and the year after Einstein's on the special theory of relativity, Oldham put out a journal article for which he continues to be remembered: the first description of the internal structure of the Earth based on measured observations.

Oldham's revelation was to be able to look at the seismographs of the Assam earthquake and separate out two different types of waves—P waves and S waves. Then, he could see that they had traveled at different speeds. P, or primary, waves are the fast ones, moving at thirty times the velocity of sound. S, or secondary, waves are slower. Not only that, but Oldham found that some of the waves had passed straight from one side of the Earth to the other, while others had taken a detour. The only way he could make sense of what he saw in the waves was to deduce that the Earth had a core made of a different material from its surroundings, and that the different material was changing the path of the waves.

At that time, six competing theories about the structure of the inner Earth were in play among geologists, mathematicians, and physicists. Oldham's finding torpedoed five of them. The inner Earth theorists fell into the same two eighteenth-century factions that had fought over the source of the magma that had erupted from the volcanoes of the Chaîne des Puys in France: the Vulcanists and the Neptunists. At that time the Vulcanists believed that the crust of the Earth had been formed by heat, while the Neptunists thought it was the result of Noah's flood covering the Earth. While their passionate disagreement ostensibly centered on fire versus water, it really came down to a dispute over how old the Earth was. And that was really about when the Earth would end.

And while the Ussher calculation that the Earth had been born in 4004 BCE was losing credibility as evidence stacked up about the planet's far older origin, the primary geological textbook for both camps was still the Bible. Geology was theology. That's why people thought the planet's time of death was inextricably tied to its time of birth. The Old Testament Book of Genesis was their particular guide to interpret their findings in the planet's rock record. By the late nineteenth century, the two groups of theorists had turned their attention to the structure of the inner Earth and had become known as the solidists and the fluidists. Among the players in this pitched century-long discussion were Ampère, Davy, and the Irish-Scottish physicist William Thomson, who became Lord Kelvin; the scale of absolute temperature measurement is named after him. A disciple of Faraday, Thomson died in 1907.

Some of the fluidists were convinced that the Earth was filled with a central primitive heat that had melted everything inside, leaving only a thin crust overtop. In that model, volcanoes and earthquakes were a direct conduit to the seething cauldron below. Others in the same bloc said the crust was thick, but still enclosed a bubbling liquid that was a by-product of the formation of the Earth, an "ejectum from the solar furnace." Still another analysis was that the planet held all three states of matter. Deep inside it was gas, surrounded by liquid and then crusted over by a solid.

And then there was the camp of the hard-boiled egg: The Earth was solid from core to crust. This theory's most famous proponent was Thomson, who declared that the whole planet must be tougher than steel and immovable within. Otherwise, he argued, "its figure must yield to the distorting forces of the moon and sun." In other words, any liquid within the Earth would be shaped by the violence of tides, just as the ocean was, throwing the planet out of balance throughout the course of every day. Thomson even delivered a lecture in Baltimore in 1884 on the topic, in which he twirled a raw egg and a hard-boiled one to demonstrate his theory. The raw egg wobbled a great deal; the hard-boiled spun like the Earth. It was good theater, if questionable science. A variation on Thomson's theory was that the Earth was very nearly solid, with a thin liquid layer just under the crust.

The sixth idea was that the Earth had a thick crust, liquid interior, and solid core. This was closest to what Oldham's interpretation of the Assam earthquake seismological data supported. The other theories soon withered. The fallout was akin to that from J. J. Thomson's discovery in 1897 of the electron, the first subatomic particle, for which he got the Nobel Prize in 1906, leading to widespread adoption of the theory of atomism, even by previous skeptics, and Bohr's model of the atom.

The magnitude of Oldham's finding is difficult to overstate. Geophysics had evolved from Aristotle's idea that the Earth was an immutable dullard in a glorious heaven, to Gilbert's contention that the Earth had a magnetic soul, to the surprising finding in the English garden of John Welles in 1634 that the magnetic field was constantly on the move, to Brunhes's finding that the entire direction of the field had switched at least once, to Oldham's charting of the shape of part of the Earth's interior. Oldham's finding began to get at the heart of why the magnetic field was so kaleidoscopic. This new information held out the possibility that the magnetic signals that people had been measuring for hundreds of years were a proxy for the architecture and even the strategy locked within that hidden place. Seismometers could finally pierce the crust, allowing scientists to peer within the heart of the planet for the first time. The key was to understand that the speed and direction of the waves contained information about the chemical composition and state of the matter they were traveling through.

Around the time that Oldham was working up his 1906 paper, Inge Lehmann felt her first earthquake. She was in her teens, she recalled, at home in Copenhagen, when the lamp swayed and the floor began to move. She didn't reveal whether that earthquake, whose epicenter was never discovered, sparked her lifelong love affair with seismic waves. But three decades after Oldham's great revelation, she published one of the most important discoveries ever made about the composition of the Earth's deep interior and therefore about how our planet came to be. The Cambridge physicist Sir Harold Jeffreys had already concluded, in 1929, that because S waves could not pass through the core, it must be completely fluid. It was the first evidence to support Halley's idea from the late seventeenth century that the core was liquid. It was a huge breakthrough and richly symbolic: The underworld of myth and Old Testament was now laid bare. Jeffreys, writing to Lehmann, then a seismologist in Copenhagen, about the reaction of his American colleague, the Jesuit priest James Macelwane, said: "I should have thought a good Jesuit would have jumped at the discovery of hell, but he reacts all wrong."

But then Lehmann took a closer look at the seismic shocks that traveled through the Earth and saw a slightly different picture. There was a discrepancy in the waves that could only be explained if Jeffreys's liquid core had another core nestled within it, somehow different from what surrounded it. Famously, the name of Lehmann's 1936 paper that explained the idea was simply "P'," after the P waves her seismometer was reading. (P' represents the type of P wave that passes through the mantle into the core and then into the mantle again.)

The tale of Lehmann's discovery is another confluence of unlikely events that characterize so much of the exploration of the Earth's electromagnetic field and interior. The only woman in the emerging international field of seismology, she was born in 1888, the child of an eminent Danish family that included artists, politicians, scientists, and a surgeon. Her father, Alfred, was a professor of psychology at the University of Copenhagen who launched the practice of experimental psychology in Denmark. He was so immersed in his work that his family only saw him when they ate together and occasionally when he took them on walks on Sundays. Lehmann's parents sent her to one of the first coeducational schools in Denmark, run by Hanna Adler, whose sister was Niels Bohr's mother. Bohr, who was three years older than Lehmann, occasionally taught there. Adler, one of the first women to get a university degree in physics, famously traveled around the United States gaining entrée into the best society by trading on her ability to explain Maxwell's equations, then new. It was similar to the tack Ørsted had taken to get into influential company in Europe at the beginning of the 1800s by carting around a brand-new voltaic pile.

In what turned out to be a boon for the field of geophysics, Adler not only believed in educating girls and boys together, but she also believed in treating them as equals. Each of her students, male and female, studied academic subjects as well as woodworking, soccer, and needlepoint. Lehmann, who died at age 104 in 1993, wrote later in life that Adler recognized no difference in the intellectual ability of boys and girls. Neither did the teachers she hired. Lehmann loved mathematics, and as a treat, her maths teacher gave her tougher problems to solve, much to her parents' dismay. They felt she was too weak to take on the extra work. Lehmann later wrote that she had simply been bored.

Lehmann hit up hard against a different philosophy from Adler's when she entered Newnham College at the University of Cambridge in 1910 after a stint at the University of Copenhagen. At Cambridge, Lehmann experienced "severe restrictions" on her movements as a woman, she later wrote, "restrictions completely foreign to a girl who had moved freely amongst boys and young men at home." And while Newnham College was established for female students, the university itself didn't allow women to earn degrees until 1948, Cambridge being the last university in the United Kingdom to do so.

Lehmann had a breakdown in 1911, which has been put down to too much work. She returned to Copenhagen, honing her mathematical skills in an actuarial office, where she calculated risks of death for insurance policies. At the age of thirty-two, she finally got the equivalent of an advanced degree in physical sciences and mathematics from the University of Copenhagen. She remained in actuarial work for a couple of years until she happened on the geophysicist Niels Erik Nørlund, director of the Danish geodetic institution Den Danske Gradmaaling. (Nørlund was married to the sister of Niels Bohr. Denmark's intelligentsia was small and well connected, then as now, and Bohr, the superstar physicist, was the node around which some of it rotated.) Nørlund recognized Lehmann's mathematical genius and in 1925 asked her to become his assistant and to set up a network of seismological observatories in Denmark and Greenland. She had never seen a seismograph before and taught herself how to interpret its squiggles before finally being sent on a three-month training trip to study with European experts. Seismology became her passion.

When she became chief of the seismological department of the Danish geodetic institution in 1928, she was in charge of interpreting the data from her seismographs and writing up bulletins. Throughout her twenty-five years in that position, she ran the office alone, rarely even having secretarial help. One of the bugaboos of the job was making sure that the Scoresbysund seismological station she had set up in northeastern Greenland was kept staffed. It was so remote that its keeper had contact with the home office just once a year, when a boat showed up. Keepers kept quitting. As for scientific research, that was not part of Lehmann's job description and was not encouraged. But it was tolerated.

This was no barrier to Lehmann. She was famous for her limitless ability for hard work and for her irritable intelligence. A relative recalls her telling him: "You should know how many incompetent men I had to compete with—in vain." And she was tenacious, perhaps imperious. A colleague recalled that she was extraordinarily sensitive to noise—another type of wave—and once, at a conference in Zurich with him, persuaded him to swap his quiet downscale hotel room for her expensive one because, despite the cost, her hotel management couldn't guarantee her that it would be quiet. At age 102, mainly blind but professionally active, she was still going to her summer cottage in Holte, on the outskirts of Copenhagen: "Of course I am in the summerhouse," she said, offended, when a telephone caller was surprised to find her there.

She insisted that seismograms from different stations be read by the same person, giving a single person a way to track a pulse of waves from station to station. And all the while, she was perfecting what a later seismologist called "a black art": the ability to listen to the story the shock waves traveling through the Earth were telling her.

Then, on June 7, 1929, an earthquake with the magnitude of 7.8 struck near the small town of Murchison on New Zealand's south island. Lehmann's network of observatories registered some P waves in part of the Earth's interior where they were not expected. She made a bold conjectural leap: What if there were something else inside the liquid core Jeffreys had discovered, something through which waves might travel faster than in the rest of the core?

This was before the time of computers. The calculations to test a theory like hers were done by hand. Lehmann didn't even have an assistant. Her cousin's son, Nils Groes, witnessed her technique. One summer Sunday he sat with her in her garden in Copenhagen, watching her sort through cards organized in cardboard oatmeal boxes on a table she had set up on the lawn. Contained on the cards was information about earthquake times, the shapes of the waves they produced, and their velocity. Her conclusion once she'd crunched the numbers after the New Zealand earthquake was that the Earth had a second core nestled within the fluid one. It was a stunning find, missed by all the eminent physicists of the day. Ever cautious, she did not declare that the new part of the Earth was solid, just that it was different. She was such a superb mathematician that she calculated the inner core's radius at nearly what today's accepted measurement is: 1,215 kilometers. She called it the "inner" core and promptly wrote to Jeffreys, the king of seismology, to tell him what she'd found—and what he'd missed. He fobbed her off. For four years. Finally, tired of waiting for him to take a look at her data, she published her famous "P'" paper in 1936. Many of the world's geophysical luminaries accepted the idea immediately, but it took Jeffreys a few years. By 1947, it was included in seismological textbooks.

Lehmann's finding, and subsequent ones by other researchers that the inner core is solid and that the whole of the core is mainly iron, underpins the development of today's theory of the geomagnetic field. Seismology remains a critical piece of the scientific efforts to look inside the inner Earth, tracking ever finer details about its architecture, topography, and chemistry. It took up a whole session at the Nantes conference, where seismologists minutely parsed, for example, findings on two big odd zones toward the bottom of the mantle underneath the Atlantic and the Pacific. These zones seem to have sharp edges and may be chemically distinct from the rest of the mantle. Seismic readings suggest they may be made of among the most primordial stuff in the core.

Lehmann, who retired in 1953, became even more prolific after she could stop chasing keepers in Greenland, often traveling to the United States and Canada to collaborate with colleagues. In 1962, Jeffreys wrote to Bohr, asking whether she had ever been recognized for scientific excellence in Denmark. Bohr wrote to Nørlund—his brother-in-law and Lehmann's former boss—recommending that she receive the gold medal of the Danish Academy of Sciences and Letters. She got it in 1965. More than twenty years later, when she was ninety-nine, Lehmann wrote her final scientific paper. That was just as British and American physicists were learning how to read another set of clues about the inner Earth: the satellite images that could examine what was going on at the boundary between the top of Jeffreys's liquid core and the bottom of the mantle. But rather than the surprise of a previously unknown architecture, the satellite images were showing the contortions over time of the seat of the Earth's magnetic power: the molten liquid with its long-limbed gyre and warring factions. Those movements, in turn, determine how strong the Earth's magnetic shield is and whether the poles are gearing up for a move.

The whole idea that they may be poised to switch again is a far reach from what Brunhes announced in his 1906 paper. His conclusion then was that the poles had at one time been on opposite sides of the planet from where he knew them to be in that year. He refused to go further, saying it was too early to make any attempts to figure out when the reversal had happened. But more than one reversal? Reversals that seem somehow to be a critical component of the dynamo at the heart of the Earth? Reversals that could affect life as we know it? Yikes!

pharaohs, fairies, and
a tar-paper shack

Inside the conference center's auditorium in Nantes, the scientists were struggling. Not with the new findings on the inner workings of the Earth. Or with the elegant maths that described them. But with the tiny font size of the print on the screen at the front of the cavernous room. Some of the conference participants were covertly bringing out binoculars. Others were taking photographs with their iPhones and then zooming in on the information using their touchscreens. This is the mind of the scientist: If there are barriers to getting data, you figure out your own way of leaping over them. It was that instinct that drove the scientific community after Brunhes's paper on reversals in 1906. Skepticism reigned. Had the planet's magnetic field really reversed direction? And if so, how can we be sure?

Theoretical questions hinged on whether such a dramatic perturbation of the poles was even possible. If it was, what was the mechanism? What was the purpose of a reversal? Could it have happened more than once? Could it even be a recurring feature of the planet's magnetic landscape?

The practical questions were no less pressing. What if Brunhes's terracotta didn't mean what he thought it meant? Modern inquisitors might question whether Brunhes's finding meant that the poles were stable but the European continent had rotated 180 degrees on the Earth's surface. But at the beginning of the twentieth century, most geologists thought the continents were fixed, so they were looking for a different explanation for Brunhes's discovery. What if scientists had a faulty understanding of the way the magnetic memory of rocks worked? What if Brunhes's chunk of rock had merely been hit by lightning and that had shifted its dip? What if rocks could change their magnetic memory on their own, without any influence from the Earth's poles?

It was this latter issue that dominated magnetic studies for decades after Brunhes's paper. If a rock could spontaneously change its record of magnetic coordinates, then the whole idea that the field had reversed would be in question, as would many other aspects of rock magnetism. Before that question could be settled, other findings supporting Brunhes began trickling in from other parts of the world. Scientists had begun their meticulous job of collecting new data points. The most compelling findings were in a modest three-page paper in *Proceedings of the Imperial Academy* by the Japanese geologist Motonori Matuyama in 1929. A professor at Kyoto University, Matuyama also studied at the University of Chicago.

Japan is a global volcano hot spot. It sits at the juncture of four tectonic plates along what's known as the Pacific Ring of Fire. Recently analyzed undersea marine sediments show that volcanoes have been active in the area for 10 million years and that the past 2 million years have been a period of extreme volcanic activity. In other words, the Japanese are keenly interested in what happens under the Earth's crust, and that island nation has produced some of the world's most eminent experts in all things inner-Earth. Including lava.

In theory, as Melloni and then Brunhes had reasoned, lava would take on a record of the intensity and direction of the magnetic field from the time and place where it cooled, in effect becoming a sort of sophisticated fossil compass showing inclination, declination, and field strength. So in 1926, Matuyama went looking for ancient basalt in a cave celebrated for that type of rock in Japan. He carefully measured its magnetic coordinates while it was in the cave and then took a sample for later examination. Its field pointed in the exact opposite direction from where the Earth's field pointed in 1926. Matuyama then embarked on a systematic examination of basalts that had spewed forth from volcanoes over many millions of years in Japan, Korea, and Northeast China, then called Manchuria. His findings were that some of the rocks' fields were aligned with today's north and some of them were aligned with the south. Few were aligned anywhere in between.

The south-aligned rocks were from different geological periods: Some were Miocene, meaning they were as much as 23 million years old. Some were from the Quaternary, making them as much as 2.6 million years old. His conclusions were staggering. Not only was there more proof that the poles had reversed, but now there was evidence of more than one reversal. Each appeared to have lasted for a long period. Even more astonishing, Matuyama could put rough dates on when some of those reversals had taken place. All of a sudden, it looked as though geologists might be able to make a clock going back over the Earth's distant past, describing where the poles had been during each era. It was a new way of seeing the planet, akin to the first maps Edmond Halley had produced showing the wavy contours of declination across the Atlantic Ocean.

One hitch in this analysis was the possibility that rocks could change their own magnetic memory. Through the 1930s and 1940s, this was an intractable problem. Rocks were tricky. Even iron, the standard material for compass needles, could lose its magnetic sensitivity. That's why sailors in centuries past had carried a lodestone as a "keeper" to keep the iron magnetized. They would stroke it across the compass's needle every now and again to remagnetize the iron. Geophysicists handled the confusion over spontaneous rock reversals by ignoring pole reversals until they could get more data, a phenomenon the American geophysicist Allan Cox and his colleagues later put down to "the embarrassing lack, even at so late a date, of a theory adequate to account for the present geomagnetic field, let alone reversed magnetic fields which may or may not have existed earlier in the earth's history."

One clue to the solution came from the work of Louis Néel, once offered a job at the observatory in Clermont-Ferrand, where Brunhes had worked. Néel eventually went to Grenoble, where he set up that university's world-famous geophysics program. That's where Philippe Cardin worked, who gave the public lecture at the conference in Nantes. But in 1931 as Néel was considering a position in Clermont-Ferrand, Brunhes's legacy in magnetism was on his mind. So was the mystery of precisely how and why a rock retained its magnetic memory. Taking a page out of quantum mechanics, Néel began to question whether every molecule in a substance was magnetized in precisely the same way. What if there were differences? In a series of discoveries that won him the Nobel Prize in 1970, Néel found that there were. In the years following the Second World War, he advanced the concept of ferromagnetism, and, in 1949, discovered ferrimagnetism, which is a related but slightly different phenomenon. In doing so, it's said that Néel took the magic out of magnetism, because he could finally explain why a material could hold its magnetic charge.

The reason goes back to the unpaired spinning electron.

The electron's motion makes a tiny circulating current. That, in turn, creates a magnetic field with two poles. In most materials that make up our universe, the magnetic fields of unpaired spinning electrons cancel each other out, so the material doesn't hold a magnetic charge. It's a nano zero-sum game. It's why so few materials retain magnetization over time. Sometimes, though, when electrons are unpaired, they don't cancel out but reinforce each other by lining up. It's the opposite of what you'd expect, and that makes these substances odd. When the electrons line up rather than neutralizing each other, the material ends up being magnetized, either for a while or, sometimes, permanently—as long as it doesn't get heated up past its Curie point. The permanent type is called remanent magnetism, after the Latin word for "remaining." This can get a lot more complicated. Even the *Encyclopedia of Geomagnetism and Paleomagnetism* says there are too many types of remanent magnetism for it to review. The type we're interested in here is the magnetism a rock acquires under natural conditions as it cools. It's commonly called natural remanent magnetism. In the days since Brunhes, scientists have learned how to strip away from rocks little bits of magnetism that came from other outside sources in order to reveal natural remanent magnetism. Carlo Laj, a French geophysicist, went back to Pont Farin and redid Brunhes's experiments after stripping away extraneous magnetic influences. His paper, published in 2002, showed that Brunhes's findings were absolutely correct.

Néel found that there are crucial differences in how the electrons decide to arrange themselves in order to enhance their magnetic fields. The differences determine the tenacity of a material's magnetic field. In some substances the orbitals where electrons live overlap across atoms. And in some of those cases, when orbitals overlap, the electrons in adjacent atoms are then forced to line up in the same direction. That magnifies the magnetic pull of a material. When that happens, the material is called "ferromagnetic." The common ferromagnets are iron, nickel, and cobalt and some compounds they are in. The name comes from the Latin word for iron: *ferrum*. The iron in a compass needle is a ferromagnet.

There's a catch, though. The enhanced magnetic field within the groups of atoms or molecules is confined to domains, or neighbor-hoods, within a material. And while the field is strong within that neighborhood, it can be offset by an opposite field in the next neigh-borhood. So the material as a whole is not necessarily magnetized. That's why your car keys aren't magnetized, as a rule. But ferromag-netic materials can be magnetized if you put them in the presence of a strong magnet. The magnet's power can make the unpaired electrons spin in the same direction, no matter which domain they're in. That's how the keepers kept the compass needles working. Stroking the nee-dle with the lodestone made the domains line up. Ferromagnets can keep this strong magnetic charge for a time, but not permanently.

And then there's more permanent magnetism, like the lodestone. Sometimes the way the atoms line up means that the opposite spins of the electrons don't fully cancel each other out. Instead, they line up in, you could say, teams of uneven sizes, in alternating rows. One team is spinning in one direction. The other team spins in the opposite. The direction of spin of the bigger team wins out for the material as a whole and the material locks in on its magnetic direction. This is called ferri-magnetism. This arrangement of spins is far more stable than that of the ferromagnetic materials. It's less apt to be changed or lost. The best example on Earth is the lodestone, the same magnetite that Homer wrote about and that Gilbert experimented on and that first sparked human investigation of magnetism. Magnetite is a type of iron oxide made up of a molecule of three iron atoms—each with four unpaired spinning electrons—connected by four oxygen atoms. It can hold its magnetic charge for millions of years, unless it is heated up past its Cu-rie point. Some rare earth elements are also ferrimagnetic.

Once Néel worked out the difference between the ferros and ferris—in my giddier moments I call them the pharaohs and the fairies—he looked at fine-grained volcanic rocks and found that they often contained enough of the right size of ferrimagnetic grains of iron oxides to bind their magnetic memory for millions of years, unless heated. The same phenomenon holds true in some types of sedimentary rocks, like iron-rich terracottas.

During the same period after the Second World War, John Graham, a keen young geology graduate student at the Carnegie Institute of Washington, DC, launched a series of expeditions to test the magnetism of rocks across the United States. Pictures from the era show a truck made into a roving rock-sampling lab, complete with a spare tire strapped to the hood. Disconcertingly, in light of the European and Asian findings, Graham found rocks in the same layer that seemed to be pointing magnetically in different directions. Could they have reversed themselves spontaneously?

He turned to Néel. Néel, a theoretician, predicted that it was possible and set out several rare scenarios in which it could happen. Supporting his theory, Japanese scientists showed in the laboratory that some lavas from Mount Haruna were susceptible to reversing their own fields, as long as they were cooled at a specific rate and contained a specific chemical composition. Yet Jan Hospers, a graduate student at Cambridge, who examined layers of lava flows from the highly volcanic Iceland, found clear evidence of not just one or two but three reversals of the whole field over time. He concluded in 1951 that "the earth's magnetization has suffered repeated reversals, and that rock magnetism can be used for geological correlation. . . .".

Back and forth it went. The Earth's magnetic rock record was reliable. The Earth's magnetic rock record was tainted. And this was the only tool the geophysicists knew of to determine whether the poles had reversed. They were flummoxed. Dueling theories about how rocks' orientations were laid down continued for years, even as evidence mounted from around the world that the Earth's magnetic field had reversed many times. By 1963, a poll of twenty-eight leading paleomagnetic researchers attending a meeting in Munich found only half could support the idea that the poles had switched places, but each of them believed that some rocks could switch their own magnetic field independently.

A year later, in a development that marked a shift in the research away from Europe and toward the Americas, there was more evidence than ever that reversals were part of the planet's inner makeup. In 1964, Allan Cox, Richard Doell, and Brent Dalrymple of the US Geological Survey in Menlo Park, California, published their landmark paper "Reversals of the Earth's Magnetic Field" in *Science*. They had gone looking for the ultimate proof that the rocks could tell the story of what had gone on in the Earth's core. That meant they needed rocks showing magnetic memory from the same time periods at different places across the Earth's crust. And that meant knowing with a high degree of precision how old the rocks were. They used a new technique involving the radioactive decay of potassium-40 to argon-40. Today, it's known as K-Ar dating, after the chemical symbols of the elements. (K is for potassium and Ar is for argon.) By determining how much argon-40 is in a sample compared to the radioactive potassium-40 it would have started with, you can tell how long it's been since the rock crystallized.

With the help of scientists around the world, they collected sixty-four samples of volcanic rocks from North America, including Hawaii; Europe; and Africa, and analyzed their ages using the K-Ar dating method. At the same time, they looked at the rocks' magnetic signals. The Geological Survey gave them a small tar-paper shack, where they could work out what it all meant. Their results produced nothing less than the first global magnetic calendar going back 4 million years, describing epochs in the Earth's history when the poles had been where they are today and others when they had been on opposite sides of the planet. Those early findings already showed some of the peculiarities that we now know characterize reversals: They last a long time in geological terms, and long enough to be captured in the rock record; they are of irregular length; sometimes, the poles try to reverse but fail.

Most intriguingly, Cox's group pinpointed the last time the poles reversed to 780,000 years ago. That's before our species, modern humans, was on the Earth. And they decided to name the current epoch in honor of Brunhes. The epoch that preceded this one is called the Matuyama. Others are named after Gauss and Gilbert. Nearly sixty years after his paper on the terracottas of Pont Farin, Brunhes's contribution to the discipline of geomagnetism was formally acknowledged.

While Cox, Doell, and Dalrymple dealt at length in their paper with the idea that rocks could spontaneously reverse their polarity, they concluded that such events were rare. In fact, they were so rare that they did not negate the robust evidence from around the world supporting the switching of the poles.

The poles really do switch places sometimes.

At last, another element of the Earth's turbulent past was starting to come into focus. Now, what the magnetic researchers wanted was to piece together reversals going even further back in time, hopefully back to the birth of the planet's own magnetic field when the Earth was about 1 billion years old, or perhaps even younger. How often had the poles flipped, and were they conspiring to do so again?

zebra skins under the sea

Any disagreements over rock magnetism through the first decades of the twentieth century paled in comparison to the most anguished geological question of that era: Do the continents move?

The idea had roots in the early nineteenth century, when Alexander von Humboldt, who traveled extensively in South America, noted how that continent's eastern flank would nestle neatly under Africa's west shoulder if the two were put together. By 1912, the German geophysicist and meteorologist Alfred Wegener gave two public talks that took the idea much further. He proposed that all the continents had once fitted together like a giant jigsaw puzzle, making a supercontinent that had later pulled apart. The Earth's crust was not fixed; it was malleable. Wegener named the supercontinent Pangaea, after the Greek words *pan* and *Gaia*, or "all Mother Earth." As evidence, he pointed not only to the shapes of the continents, but also to similarities across the modern continents of geological features and species. He even drew a map, which was later disparaged for its lack of precision, showing where the continents might once have been when they formed Pangaea.

Laid up by an injury in the First World War, Wegener refined his ideas and published a book on them in 1915. It was a scientific scandal. Wegener was criticized for the scientific no-no of being a proselytizer for his unorthodox and unpopular idea, which became known as "continental drift." Protocol demanded that he seek truth, not converts, his critics said. Ostracized, he could not find work at a university in his home country, and finally took a position in Austria. Fifteen years after he published his book, and long before the scandal abated, he died, trapped in a storm while trying to ferry supplies to a meteorological station in Greenland by dogsled. He was fifty.

Why was it such a flash point? Cambridge's Sir Edward Bullard, one of the British lions of geophysics, who first repudiated Wegener's ideas and then championed them, wrote about the backlash in a retrospective essay in the 1970s. "There is always a strong inclination for a body of professionals to oppose an unorthodox view. Such a group has a considerable investment in orthodoxy: they have learned to interpret a large body of data in terms of the old view, and they have prepared lectures and perhaps written books with the old background. To think the whole subject through again when one is no longer young is not easy and involves admitting a partially misspent youth." Until the 1950s, believing in Wegener's theory of continental drift was "unusual and a little reprehensible," Bullard wrote.

But then a set of clues supporting the idea began to emerge, rather haltingly. In the early 1950s, Edward "Ted" Irving began working on his PhD in geophysics at Cambridge, focusing on rock magnetism. One of his fellow students was Jan Hospers, who had looked at the Icelandic lavas and had seen reversals. Irving began to study a magnificent stretch of exposed sandstone in the northwest of Scotland. The Torridonian sandstones were hues of red or purple or brown and they ran horizontally for seventy miles along coastal mountains in beds at times 18,000 feet thick. They were untouched, as much as 700 million years old, fine-grained, sprinkled with magnetite and hematite, and strongly magnetized. Irving took four hundred samples.

But when he analyzed them, he found that their magnetic fields pointed to the northwest and southeast, far from the present-day geographic pole. He toyed with the concept that the rocks had reversed their magnetic direction on their own, but began exploring another idea, along with Kenneth Creer, a postgraduate student at Cambridge. They found that the older the rock was, the farther away from the current pole the field pointed. Could it be that the poles wandered across the face of the Earth? They plotted their "paleopoles" on a map charting the possible sites of the magnetic north pole back 700 million years. In 1954, they presented their "polar wander path" at the meeting of the British Association for the Advancement of Science. The idea caught on. An article in *Time* magazine that year rather excitedly tracked the "north pole's travels" at 14,000 miles in 700 million years, working out to 1.3 inches a year, complete with a diagram.

This was not the poles switching places during a reversal, which scientists had been considering since Brunhes's paper in 1906. This was the poles meandering far from the axis on which the planet spins in an apparently determined progression. It was different from the observations of long-term idiosyncrasies in the magnetic field, or secular variation, which saw the poles shifting around, but always in the vicinity of the geographical poles. If the idea that the poles could wander to remote parts of the planet had been true, it would have added another level of mystery to the workings of the inner Earth. Scientists didn't have a scrap of theory to explain it.

They didn't need one, as it happened. Irving and Creer never imagined that the poles were doing that type of broad-scale roaming. Even as they were charting their polar wander path, they thought it was far more likely that the poles had stayed put (more or less) and that the rocks themselves had moved, embedded in Scotland, which was moving too. They re-dubbed the phenomenon "apparent polar wander." Irving realized that what he was tracking was not the changing position of ancient poles but the changing position of ancient latitudes and therefore continents. He had read his Wegener. Using this "apparent polar wander" information, he could retrace the march of the continents over time, each compared to the others. It was like seeing lost worlds spring back to life.

To test his theory, he got blocks of basalt from seven ancient lava flows in India and found that, based on the information contained within their magnetic readings, the Indian continent had moved north by 53 degrees of latitude and rotated counterclockwise by 28 degrees from about the time the dinosaurs died out 65 million years ago. The shocking conclusion from his results, when he put them all together, was not only that the continents had moved but that they had moved immense distances over time.

His ideas were so controversial that when he wrote up his findings for his doctoral thesis, the Cambridge examiners refused to award him the degree. In 1954, Irving took himself off to the Australian National University in Canberra, had a beer for solace with his new boss, dusted himself off, and pressed on, finding ever more evidence for his theory. He ended up in Canada, drawn by his Canadian wife and the allure of the country's pre-Cambrian shield, where some of the planet's oldest rocks are exposed.

At the same time, in the years following the Second World War, another set of clues was finding its way into the mix. Geophysicists had begun a more thorough examination of the seafloor. It was a whole new discipline called marine geology, and it involved ships dragging echo sounders, dredgers, and seismographs across the ocean, as well as machines to drill cores in the seabed. It began as a military enterprise by governments wanting tactical information about the shape of the ocean floor. At that time, even some eminent geologists contended that the ocean contained a network of sunken continents that could surface almost at will. Many thought that continents had once been ocean bottoms. Most thought of the seafloor as featureless and barren. The new findings painted a different picture. They said that the bottom of the ocean was not made of the same stuff as the continents. Ocean floor was basalt. It was thinner and far younger than the continents, with nothing more than about 200 million years old. As for the ocean's hills, they were underwater volcanoes.

Beginning in 1956, some of that deep seabed topography could be visualized for the first time. A team at the Lamont Geological Laboratory at Columbia University in New York (today, it's the Lamont-Doherty Earth Observatory), under the direction of the eminent geophysicist Maurice "Doc" Ewing (he was a great friend of Inge Lehmann, and her collaborator), had begun collecting tens of thousands of deep-ocean soundings made over many years in journeys across the Atlantic Ocean. Marie Tharp, one of the mathematicians who worked on them, recalled that a neat-handed colleague recorded the data using a crow feather quill pen, writing in India ink on blue linen pages, a document that became the bible of the field. Tharp helped plot the data on physiographic maps that could show the ocean floor's profile as if it were being spotted from a low-flying airplane. The early, patchy images clearly showed a deep cleft valley slicing between a curving line of mountains on either side. When Tharp showed it to her boss, Bruce Heezen, he groaned, said it looked "too much like continental drift," and dismissed it as "girl talk." Nevertheless, Tharp persisted and by 1956 had a convincing map of the Mid-Atlantic Ridge. When Heezen and Ewing presented the map to a meeting in Toronto of the American Geophysical Union that year, scientists reacted with amazement, skepticism, and scorn, Tharp wrote. A mid-ocean ridge representing a seam in the ocean floor where new floor was being created, forcing continents to move across the face of the Earth? Silly girl.

As the 1960s dawned, it occurred to geophysicists on either side of the Atlantic Ocean that they might be able to take magnetic readings of rocks on the seafloor by towing a magnetometer behind a ship. Results began to emerge from the deep ocean, including a batch from the East Pacific Ocean Basin. Intriguingly, they showed bands of magnetic readings, fields pointing in alternating directions, right across the deep seafloor, lining up parallel to the East Pacific Ridge and symmetrical on either side of it. The results were published, but without explanation. No one could understand what they meant.

Enter the Canadian Lawrence Morley, a specialist in rock magnetism with the Geological Survey of Canada. He became obsessed with the findings on the ocean bands, admitting in a later essay that he neglected all his other duties while he tried to figure them out. He was familiar with magnetic readings over broad terrestrial landscapes because he had done them from the air to find petroleum and minerals. On land, they were a mishmash of polarities. They were not in tidy patterns like those on the ocean floor. But he was certain the ocean stripes must be related to remanent magnetism, as those on land were. Then he discovered a 1961 paper describing the evolution of the ocean basins by seafloor spreading.

In a flash, he put three concepts, until then unconnected, together into a single theory: Wegener's continental drift, seafloor spreading, and reversal of the poles. In his view, the stripes on the ocean floor came from hot magma continually rising from the inner Earth at seams in the Earth's crust, creating brand-new ocean floor. As the bands cooled in the water past their Curie point, the ferrimagnetic materials in them took on the direction of the magnetic field from that time. They would spread from the ridge, or seam, in the crust, symmetrically on either side toward the continents, laying down precise records of the magnetic field. Over millions of years, as new seafloor was born, it became an archive of pole reversals. If you colored the bands of negative polarity in black, the picture looked like a zebra skin, spreading out from a central meridian: black, white, black, white, and so on.

Morley swiftly wrote up a paper explaining his hypothesis and tried, as he said, "desperately" to get it published. *Nature* rejected it in

February 1963, saying it lacked the space to carry the article. It languished for months on the desk of an anonymous reviewer for the *Journal of Geophysical Research* before again being rejected in late August. In a note appended to the rejection, the reviewer said the idea was interesting, and added, in a remark so snide that it has been engraved in scientific memory, that it was "more appropriately discussed at a cocktail party than published in a serious scientific journal." On September 7, 1963, *Nature* published an article by the Cambridge geophysicists Frederick Vine and Drummond Matthews, who had independently come to the same conclusions as Morley, based on magnetic readings from ocean ridges in the Indian, Atlantic, and Pacific Oceans and the paper on seafloor spreading. The Cambridge team was gracious. Today, the idea is known as the Vine-Matthews-Morley hypothesis. For his part, Morley abandoned rock magnetism and became a pioneer in remote sensing with satellites.

By late 1966, there was a continental divide on the idea of continental drift. Most American geophysicists rejected it, while most Europeans accepted it. Bullard recounted arriving in New York that year at a key symposium on the history of the Earth's crust. On the first day, Ewing, whose own team had produced the first maps of the mid-ocean ridge system, said to him: "You don't believe all this rubbish, do you Teddy?"

In the meantime, Cox, Doell, and Dalrymple, who had sweated over their magnetic chronology of the Earth, took a look at the zebra stripes now being produced by magnetometers dragged over spreading ocean floor ridges. By then, geophysicists were able to calculate with mathematical models how long it had taken for a stripe to form, meaning that the patterns were not only chronicles of magnetic direction, but also of time. When Cox and his group placed the ridge readings against the timeline they had put together in their 1964 paper, the two correlated. Cox said: "I felt cold chills. This was the most exciting moment of my scientific career."

Wegener's idea, combined with the Vine-Matthews-Morley theory, Tharp's maps, Irving's work on the shifting latitudes, and Cox's magnetic clock, led to what became known in 1968 as the theory of plate tectonics. That's the idea that the Earth's crust moves slowly across the mantle on about twenty plates of different sizes, some huge, others small. Some parts of the seabed floor are spreading at ridges or rifts; others are being consumed, or subducted, underneath one another at trenches. It's the ultimate in recycling: Old floor gets destroyed along one plate boundary, and at another, new floor gets made. Moving plates can also cause continents to collide. That's what happened about 50 million years ago when India and Asia crashed into each other. Rather than one plate sinking under the other, the plates rose up to make the Himalayas. At still other boundaries, the plates move in opposite directions against each other, grating, creating fault zones that are prone to earthquakes. One of the few on land today is the San Andreas fault zone that slashes through California, more or less parallel to the coast.

By the mid 1970s, plate tectonics was a foundation of geology. Wegener's reputation was resurrected. By 1980, the Germans had named an eminent research institute after him. It focuses on the oceans and polar regions. Even Wegener's reconstruction of the existence of the vanished supercontinent of Pangaea, with some massaging and maths from Bullard, had become doctrine. Recently, using calculations from seismographic data, geophysicists have been able to reconstruct long-vanished crustal plates that were subducted into the mantle as much as 250 million years ago, like a record of a ghost Earth. Despite the wide acceptance of the theory of plate tectonics and the evidence to support it, some scientists continued to repudiate the idea. The most vociferous holdout was Sir Harold Jeffreys of Cambridge, the same distinguished theoretical geophysicist who had initially ignored Lehmann's finding that the Earth had an inner core. Jeffreys said the mechanics of moving plates were impossible. He died in 1989 at age ninety-seven, unconvinced.

Nevertheless, the theory of plate tectonics also proved the theory of pole reversals. Relying on readings from the seafloor, geophysicists have firmly established a calendar of the Earth's magnetic chronology going back 252 million years to the boundary between the Permian and Triassic geological periods. (Models taking the calendar back even further also exist, supplemented with the rock record.) That date coincides with the biggest extinction spasm in the Earth's history, when about 95 percent of the species on the planet went extinct. The trigger for the mass extinction was an influx of carbon dioxide into the atmosphere from the volcanic eruptions that created the modern-day Siberian Traps formation. Over the past 252 million years, the record shows that reversals generally happen two or three times every million years, with at least two longer periods known as "superchrons" when they did not happen at all. In the past 90 million years, reversals have steadily become more frequent. Near-reversals, or excursions, happen about ten times as often. That's when the shield wastes away to a small fraction of its usual strength, the dipole is destabilized, and the poles move erratically as far as the equator before going back home. The last excursion happened 40,000 years ago, coinciding with the extinction of Neanderthals.

Once geophysicists knew for sure that the reversals happened and when they had happened, the focus turned to trying to figure out where the field was headed.

CHAPTER 21

at the outer edge of the dynamo

Oddly, since we were in the land of the café au lait, morning coffee was hard to come by in Nantes, at least in the quantities required for alertness at a scientific conference. So Kathy Whaler and I took a detour one morning and stopped off at the bustling train station's take-out coffee stand.

Like most at the conference, Whaler, a professor of geophysics at the University of Edinburgh, was a luminary. You couldn't move through the crowd at the conference center without bumping into people who have done the defining modern-day work on the inner Earth. The level of expertise was so high that when I would ask someone a question, I almost invariably got redirected. "Check with so-and-so, who wrote a big paper on that and is standing right over there." And then, when I would find that person—the one who knew the answer better than anyone else on Earth—there was always demurring and more redirecting for another fragment of evidence from someone else.

It was the culture of a meeting like this not to be definitive. Because the discipline itself is still evolving, the meeting was less about certainties than about tracking the slow progress toward certainties. That meant it was the place for presenting careful summaries of what was agreed on, exploring new findings, detailing methodologies to figure out more and more precise ways of reading what the data were telling you, comparing contradictory interpretations. It was where you wore a sackcloth to admit errors, accepted criticism, basked in praise, defended your approach, and, maybe most important, found out what others were looking at. The focus was on the challenges that remain, from the arcane to the urgent. So, for example, are there radioactive elements in the core? What is the chemical composition of the deep mantle? How can you look into the heart of the Earth's magnetic field without interference from the thick mantle and the crust? How is the field changing over time?

Reading what's going on within the core itself was where Whaler came in. She was at Cambridge working on her doctorate under the supervision of David Gubbins when an astounding batch of new satellite data came in. These were figures from MAGSAT, the first satellite that could read the whole magnetic vector—both direction and strength—across the whole planet. Operated by the National Aeronautics and Space Administration (NASA) and the US Geological Survey, it collected data for about half a year, ending in the late spring of 1980. (An earlier set of satellites, POGO, launched from 1965 to 1969, was the first to do a general map of the field, but only looked at its strength, not its direction.)

The MAGSAT data were superb. For the first time, researchers could look at the global structure of the field. Whaler was hooked, she told me as we walked across a bridge over the Loire with our coffees to rejoin the others. All of a sudden, you could compare these precise figures from the satellite with data from modern observatories on the ground and try to reconcile the two sets of numbers. And not only the modern numbers but the whole archive. It meant putting together data stretching back to sixteenth-century sailors' measurements of declination and inclination with the nineteenth-century measurements captured by Gauss's magnetic union and Sabine's magnetic crusade. It meant combining those with the record in the rocks that Brunhes and others were beginning to read in the twentieth century and with the new findings from the seabed floor. Finally, a big picture of the field over time was emerging. It was intoxicating.

At an initial pass, Whaler, Gubbins, and others put together a 380-year record of the field and its variation over time, Gubbins explained in an article in *Scientific American* in 1989. Some of the findings were expected. At the surface of the Earth, the field looked like that of a bar magnet lying along the same axis as the one on which the planet spins. These are the endless looping lines that flow from the south magnetic pole out into space and then back into the Earth at the magnetic north pole. The denser the lines are, the stronger the field is.

Their reconstructed 380-year record also catalogued what's known as the "westward drift" in the field over the past few centuries. The idea emerged in the late seventeenth century, when Halley suspected that the field was listing to the west, for which he found evidence in London's shifting measurements of declination. Gubbins's team tested that idea by tracking the field line on the planet where declination is 0, or where the compass points to both magnetic and geographic north. Its official name is the agonic line, after the Greek phrase for "without angle." In 1700, for example, the line ran midway through the Atlantic Ocean, curved over the Gulf of Mexico, and ran straight through the Great Plains. By 2017, it had drifted so far west that it was on the Pacific side of South America, careening up through the middle of Minnesota. This was the longitudinal prime meridian that Halley and so many others had sought hundreds of years ago, believing it would bisect the Earth into two neat halves and solve the problem of navigation at sea. Gubbins's model showed that the agonic line is wildly unpredictable. Back in the early seventeenth century, for example, it ran up through Africa, looped over Norway, slid down past Greenland across the top of South America and out into the Pacific before scooting up to the North Pole via Southern California.

And then there was the strength of the field. Scientists have kept a continuous record of the field's strength since 1840, after Gauss worked out how to measure it. And while it's possible to figure out the field's strength from proxies going back further in time—like ancient terracottas and lavas and mariners' measurements—that's not considered to be as precise as direct measurements. So to geophysicists, 1840 is a key line between observed, indisputable measurements and those that are derived from other evidence. When Gubbins looked at his maps going back over time, it was absolutely clear that the dipole had declined since those first measurements in 1840. Looking at terracottas magnetized two thousand years ago during the Roman era, the team could work out a sustained and remarkable weakening from that time.

But why was it weakening? Every single reading taken until then had measured the magnetic field as it manifests itself at the surface of the Earth. But the field, as Gauss showed mathematically in 1838, is generated *within* the Earth. Between the outer edge of the Earth's core and the surface was nearly 3,000 kilometers of mantle and crust, potentially interfering with the magnetic signal from the core. What if the field looked different closer to its source?

The trick there was to work out how to strip away any magnetic interference from the crust and the thick mantle and to see what was happening to the field as close as possible to where it originated—at the outer edge of the dynamo. Gubbins and his team wanted to know more precisely what drove the Earth's magnetic power, how it had evolved, and where it was headed. They were convinced that looking at the field at the core–mantle boundary would give them some clues.

By 1985, Whaler, Gubbins, his graduate student Jeremy Bloxham (now at Harvard), and others had figured it out, at the same time as a separate group working independently at the Scripps Research Institute in California. They used the mathematical methods devised by James Clerk Maxwell in the nineteenth century to project their readings from the Earth's surface down to the bottom of the mantle, to where the mantle hugged the outer core. The Gubbins group started with the 1980 data and then reached back in time to 1777, making maps that captured both the direction of the field and its intensity at the core. The maps showed the number of field lines and where they exited and entered the core's surface. That's known as the magnetic flux over an area. Then they colored the outgoing flux shades of red depending on how intense it was, and the incoming flux, blue.

This picture of Earth's core was a revelation, a bewildering hodgepodge of swirls and colors that betrayed a far more complicated field than anyone had imagined. As Gubbins explained, if the maps had been describing a simple two-pole system, the northern portion of the image ought to have been blue and the southern, red. It would have been deepest blue at the magnetic north pole, aligned with the Earth's axis of spin, and deepest red at the magnetic south. That would reflect the fact that the lines converge at the poles, increasing intensity. In addition, the core's magnetic equator, close to the geographic equator, would have shown up as a boundary between red and blue where no flux penetrated the surface.

Instead, the map of the core showed elements of a two-pole system but also other anatomy deep within the outer core. It was like being able for the first time to see inside the human body with a magnetic resonance imaging machine, discerning the shape of the liver and heart and lungs. For one thing, while the north was mainly blue, and the south mainly red, that was not absolute. There was red where blue was expected and blue where red was expected. Also, several blobs showed up where the flux was either greater or less than expected. Two of the low-flux patches were around the poles, the opposite of what researchers would expect. And then there were the two curious patches below the southern Atlantic Ocean and Africa. They were strong, running in the opposite direction from the way the dipole would demand: in instead of out. Not only that, but the one under Africa was moving westward at the astonishing pace of about one-third of a degree of longitude every year. As for the dipole itself, the map showed what Gubbins and his team believed to be evidence of two columns of liquid spinning in the outer core, separated from the rest of the molten metal making up the outer core. Those columns of movement seemed to be supporting the dipole. And the patch moving under Africa appeared to be what was undermining the dipole.

The elements were now in place to begin to see the kinetic workings of the molten inner Earth, to glimpse its very heart pumping. The trail of discoveries to get there had hopscotched from the development of the simplest compass by the Chinese in the centuries before the common era to Gauss's proof in the nineteenth century that the Earth's magnetic force was within the planet itself, from the realization in the seventeenth-century garden near London that the field was changeable to the twentieth-century seismic-wave data proving that the core was both liquid and solid. All these pieces of the magnetic puzzle so arduously assembled over so many centuries had fallen into place to produce a series of maps that showed not just what was in the core, and how that core gyrated, but also how it gyrated over time.

Since then, new data have poured in. By 1999, the Danes had launched the Ørsted satellite, named after Hans Christian Ørsted. It remains in orbit. Originally, it captured the whole field vector, but since 2006 has been recording only intensity. The Germans launched the CHAMP satellite in 2000, which orbited for ten years until it reentered the atmosphere and burned up. The SAC-C, launched by a robust international coalition including NASA, orbited from 2000 to

2013. In 2013, the European Space Agency began the Swarm mission—a trio of satellites measuring the whole field at the same time. Together, the satellite data represent an unbroken, high-grade record of the Earth's magnetic field, as seen from above, for the better part of two decades.

By 2000, the geophysicist Andrew Jackson, another of Gubbins's graduate students, had developed what is now a widely used, more precise computer model allowing researchers to see what was going on at the boundary between the mantle and core stretching back four hundred years. Between his model and others, critical changes in the field at that boundary have become clear. The blue reversed-flux patch Gubbins and his team discovered has kept growing and has kept moving westward. In 1984, it joined forces with a similar, smaller patch underneath Antarctica. By 1997, this formation had attached itself to the field in the northern hemisphere, meaning that a massive blob of blue now runs through the red of the southern hemisphere almost from the magnetic equator to the south pole. The non-dipole part of the magnetic field is getting stronger, a dramatic shift for a single human generation.

As for the dipole's overall intensity, it is waning too. Since the critical year 1840, it has decayed by about 10 percent, measured from the surface of the Earth. And it's the part of the field that is changing most slowly—because it is the biggest—while other wily structures in the field, led by a sinuous gyre, strain ever more wildly to break free.

CHAPTER 22

anomaly to the south

Christopher Finlay talked about the gyre in the heart of the Earth as if it were alive. It was eccentric, he said. It had limbs. It twisted and stretched. The gyre's actions were coercing the Earth's magnetic dipole to decay. To me, it sounded as if this mysterious gyre were an organism secretly sucking energy from the dipole, feeding enemy factions within the core, destabilizing the regime. Far from stable and orderly, this was anarchy. It was a breathtaking peek at a covert drama taking place within the invisible magnetic force.

If anybody could say what the gyre looked like, it was Finlay. Tall and lanky, with curly brown hair and a quick smile, he was one of a small number of people on the planet with the prowess to work it out. He grew up near Belfast, Ireland, ambling around the countryside with a compass in hand, fascinated with the heroic tales of Edmond Halley and the magnetic crusade. By the time I met him, he was a geophysicist at the National Space Institute at the Technical University of Denmark in Copenhagen, the university Hans Christian Ørsted set up in 1829. Along with institutes in Potsdam and Paris, Finlay's was one of the three European scientific centers monitoring data from Swarm, the triplet of satellites now tracking the magnetic field from space. His boss in Copenhagen, Nils Olsen, was known as the dean of Swarm data. The bottom line: If you're a scientist working on the Earth's evolving magnetic field, you know the work that Finlay and Olsen are producing.

It was through Finlay that I had come to be at the meeting in Nantes. When I saw him in Copenhagen, he had told me about it, under questioning. It would, he said carefully, be highly technical. Nevertheless, I had written to the organizers and received enthusiastic permission to attend as a journalist guest. Once there, Finlay was one of my guides to the goings-on, good-naturedly pointing out people to consult and explaining basic concepts. As the conference progressed, I ended up with a couple of pages of questions at the back of my notebook under the heading: Ask Chris!

In Copenhagen, I had caught him in the midst of preparing teaching materials and getting ready for another meeting. He was on the run. Every now and then, he would jump up and dash over to his computer to show me what he was talking about. He would click through PowerPoint presentations filled with notes for the classes he taught, and pull up colored maps showing the Earth's magnetic field. As for Gubbins before him, and then Jackson, who supervised Finlay's PhD at the University of Leeds, those maps represented an important way of describing his understanding of the field and how it is changing. Above his desk, nine printed-out maps were carefully thumbtacked to a large bulletin board. Each map save one had the same color tacks on each of the four corners. A great whiteboard dominated the wall beside it, covered with mathematical formulae and calculations in tidy blue, green, and black markings.

All this work is part of a quest to understand the dynamo inside the Earth. To do that, Finlay and his colleagues have created computerized numerical simulations to see if they can replicate the one that generates our magnetic field. The idea is to understand today's field but also predict its future movements. It turns out that a key component in the model seems to be the gyre.

The best matches in the models show the gyre to be a thick swath of molten metal in the outer core, clenched between the grip of the solid inner core and that of the mantle. Mantle and inner core are locked together by gravitational forces, but convection deep in the core tends to drag the inner core to the east. To stay in balance, the gyre close to the top of the outer core is forced to move westward. That explains the westward drift of the Earth's magnetic field that Halley first noticed. Because the inner core is growing lopsidedly—cooling faster underneath Indonesia—that too puts pressure on the gyre, distorting it into eccentricity. That explains why the long-term variation in the magnetic field happens mainly on the Atlantic side of the planet, and less on the Pacific. Bad luck for all those navigators who were tracking declination and dip on well-worn crossings from Europe to North America over the centuries. Had the greater variation in the field happened on the North America–to–Asia side instead, sailors would have had an easier time of it. Perhaps the magnetic crusade and the race to understand the magnetic field would not have happened at all.

One of Finlay's papers contains a visualization of what the models say the gyre might have looked like in 2015. Colored bloodred and dark blue, the gyre pulsates in fleshy ropes, looking for all the world like viscera. One of the many revelations here is that the Earth's inner anatomy is not symmetrical. It is fundamentally askew. This is not the image of the plodding bar-magnet within the Earth that researchers held until the last century. Or the hard-boiled-egg vision of the planet that was common just over a hundred years ago. Oldham, Jeffreys, and Lehmann, whose seismographic analysis produced the first glimpses of the Earth's inner structure, would hardly recognize this stroppy and complex creature.

And in Finlay's models, it was the combination of oddities in this limb-stretching gyre that was driving several critical phenomena affecting the dipole, and therefore the magnetic field. In the models, the gyre's limbs reached up and down through the outer core in muscular columns. In the northern hemisphere, they transported magnetic flux in a balanced fashion: What went up to the pole also came down to the equator. In the southern hemisphere, things were badly off balance. A strong magnetic flow ran from southwest of Australia toward the equator, but it was not offset by a similar flow toward the south pole under South America. On the contrary, the reversed-flux patch that Gubbins and Jackson had been tracking in the south was growing both larger and stronger. It was this lack of symmetry in the south that was draining the dipole.

That was important, Finlay explained, because the magnetic anatomy within the core is already highly diverse, another revelation that would have stunned early magnetic investigators like Gilbert and Halley. In 2017, observed from about 64,000 kilometers from the Earth's center, the dipole component of the field accounted for 99.9 percent of the field's energy. At the Earth's surface, it was 93.2 percent. Where the mantle meets the outer core, it was only 38.6 percent. The rest is smaller-scale, more complex magnetic components. The numbers at the core–mantle boundary most accurately reflect what's happening inside the Earth because it is closest to where the field is produced. It means even small changes in the dipole show that the field as a whole is changing from within the core.

In addition, the reversed-flux patch on the core–mantle boundary was linked to an odd bruise of decay in the field on the Earth's surface, like a soft spot on an apple. Again, Finlay turned to his computer. He brought up a colored map to show me. It was a two-dimensional globe in vivid greens, blues, reds, and yellows, with a spate of white dots scattered across it. This was an image of the strength of the Earth's magnetic field at the surface of the planet, a descendent of the maps made since the times of the magnetic crusade, cousins of the ones

Gubbins and his team had made of the boundary between the mantle and the core in the 1980s.

Most of it was green, which corresponded to a field strength in the range of 40,000 nT (nanoTeslas). Patches near the two poles, north and south, were dark red, indicating a stronger field closer to the range of 60,000 nT. But the curious feature was the vast expanse of blue. It stretched from the eastern edge of southern Africa across the Atlantic Ocean far to the west of South America, and from the equator almost to Antarctica. It was a region of low field strength, down in the range of the 20,000s or so. On top of the blue, and in a few other places, were splotches of white dots representing episodes when satellites flying overtop the weak field had had memory failure.

This was the South Atlantic Anomaly, named for the fact that it had a weirdly low field strength and was centered in the Atlantic Ocean below the equator. In fact, the field was so depleted that solar radiation reached close enough to the Earth's surface to disrupt satellite technology.

Its existence came as a surprise to geophysicists. They were unaware of its impact before satellites began to transmit information, partly because there had been relatively little measuring of the magnetic field in the southern hemisphere over time. They caught the first glimpses of how it was changing with the onset of the Ørsted and CHAMP satellite data in the early 2000s. But it was when the Swarm trio started transmitting far more precise information beginning in 2015 that its influence became clear. Not only is the anomaly big, it is heading to the west, growing rapidly, and causing the field to weaken fast.

And fast not just in geological terms, but in human terms too. A paper published in 2016 found that if you define the anomaly as having a field strength below 32,000 nT, its area grew by more than half—53 percent—from 1955 to 2015. By then, it covered a little more than one-fifth of the Earth's surface—20.3 percent—up from 13.3 percent sixty years earlier. In that same period, the very weakest part of the anomaly degraded fast too, dropping from 24,000 nT to 22,500 nT, a decline of 6.7 percent.

So then the question emerged: Was this twisting gyre, linked to an off-balance dipole that in turn produced the South Atlantic Anomaly, the long-sought mechanism for the reversal of the poles? Was it evidence that the inner Earth's warring magnetic tribes were in the throes of toppling the dominant dipole even now? Are the poles poised to switch places?

Finlay was constitutionally unable to tell anything but the whole truth. That was the alpha and omega of his scientific training. He would love to be able to say for sure that the gyre is the key to a reversal—or that it isn't. He would love to be able to say whether a reversal is on its way right now. But he can't. The most definitive thing he can say is that the poles will certainly reverse at some point, as they have so many other times, and that we cannot rule out that this could be the early stages of a reversal.

The truth is that neither he nor anyone else knows what the start of a reversal looks like. There are theories. There is no consensus. There's little detailed evidence in the rocks of precisely what was happening in the core when the earlier reversals took place. In geological terms, the reversals happen too quickly to lay down a good record of the transition from one field direction to the next. In fact, it's not clear that rocks can always capture a magnetic signal in a highly disturbed, reversing field. Often they are only able to record the fact of the reversal itself.

There's not even a consensus on what causes a reversal. Maybe it's the gyre. Maybe it's something else. When Finlay and others make their computer simulations produce a reversal in the field, sometimes it resembles what's happening today. A patch of inverted polarity can grow and move from equator to pole and then—a flip. But in some of the simulations, the reversed-flux patches grow powerful and then the dipole beats them back and regains control. The reversal is averted.

And if this is a reversal, how advanced might it be? That depends where we are in the process and how long the process takes. There's no consensus on any of it. A reversal has three separate phases. There's the period when the dipole is weakening, followed by the relatively swift movement of the poles to opposite sides of the planet, followed by the regrowing of the dipole. No one knows exactly how long each of these phases lasts, or even whether they are consistent during each reversal. The general thinking is that each phase is at least centuries long, perhaps longer, and that anything quicker is implausible. Even that is still under dispute. A recent paper by the Italian researcher Leonardo Sagnotti that looked at a continuous layer of sediments in the Apennine Mountains calculated that the last time the poles reversed, 780,000 years ago, it happened in less than a century. Most other evidence suggests it took about 10,000 years from beginning to end. This is a crucial point because the concern is not primarily the reversal itself. It is the weakening of the field while the reversal is in process and how much additional radiation will strafe how close to the Earth for how long.

And while the issue of how long the dipole has already been weakening is contentious, geophysicists can say that today, it is about twice as strong as it was just before the five previous reversals. It's been decaying by an average of 16 nT a year since the first Gaussian measurements of the field's intensity around 1840, which is less than two hundred years. That's a total of about 10 percent since 1840. And that's on the surface, not at the core–mantle boundary. If the whole dipole were to continue to decay at that rate, that part of the field would be gone in less than two thousand years. But it's unclear just how far the dipole would have to decay in order for non-dipole factions to force a pole reversal. Would it have to go all the way to zero? Or something short of that?

And, as Finlay pointed out repeatedly, the magnetic field does not function according to a linear model. It is profoundly, intrinsically nonlinear. To physicists and mathematicians, nonlinearity has a precise meaning. Linear means you can add up all the components of something and get a correct answer. So, if I'm making a cake and I double the ingredients, I get twice as much cake. Nonlinear means the answer isn't directly proportional to the sum of the components that go into the problem. If each component doubles, the endpoint is not necessarily double. So you can solve individual parts of a problem, but when you put them together, you might not get the answer you expect. Not only that, but when components of this nonlinear system are changing, it gets even harder to figure out what the answer will be.

And then there's chaos. Some nonlinear systems are also chaotic, which means that a tiny change in initial conditions can have a profound, unpredictable, and counterintuitive change in outcome. Just because something acted a certain way before doesn't mean it will act that way in the future. It doesn't mean random; these systems still follow well-defined laws. Also, a chaotic system doesn't show a discernable pattern over time. The most famous way of explaining the idea of chaos comes from the world of meteorology. Trying to work out a way to predict weather in 1961, the American mathematician and meteorologist Edward Lorenz ran a computer simulation and then reran it, starting it in the middle of a time sequence. But he inadvertently truncated one of the numbers he input by a few decimal places. The computer program, which didn't change, produced a dramatically different forecast. Lorenz eventually described it this way: If a butterfly flaps its wings in Brazil, does that set off a tornado in Texas? His explanation became known as the "butterfly effect." Small changes can lead to big differences.

I asked Finlay if the core is chaotic. Maybe, he said. He paused and thought for a moment. The core is a turbulent place with strongly nonlinear dynamics. The reversals don't show a simple pattern in time. The models show it is sensitive to initial conditions. Does that make it chaotic? It is certainly a very plausible hypothesis, he said.

The idea of nonlinear chaos has much deeper historical roots than the 1960s, though. For more than 250 years, ever since Isaac Newton presented his theory of gravity, mathematicians have been trying to solve something called the three-body problem. It goes like this: You have three particles (or heavenly bodies, originally) moving in space, connected by their gravitational pulls. You know where they are now. Tell me exactly where they will be in the future. It turns out to be unsolvable except in a few weird scenarios. You can solve it for one body and for two, but not for three. It means the movement of the bodies is nonlinear and therefore impossible to predict over time.

It's similar in the core. You can know roughly what the core is doing now. You can know more or less what it did in the past. You can know the rules of physics the core must abide by and that they say the field's direction must change at some point. But you can't say for sure when it will happen. The nut of the problem is precisely that difficulty in knowing the initial conditions. Any tiny changes make for enormous differences. Added to that, the reversals are aperiodic, meaning, unlike in the sun, they don't happen in a time pattern that anyone can discern.

This business of trying to predict the field's future has been going on in some form or another since Henry Gellibrand's measurements in John Welles's garden in 1634. It has been primarily theoretical and mainly carried on in the absence of enough knowledge or data. But the efforts based on more precise information are far more recent— only a few decades old. The Swarm data, which are the most precise of the lot and which allow for more detailed calculations, only began after the satellites were launched in late 2013. This is the cutting edge.

Here's a measure of how much of a frontier these scientists are on. One of the first things I noticed in Finlay's office in Copenhagen was a full-color replica of the map Halley made of declination lines across the Atlantic Ocean, based on his observations on the *Paramore* in 1700. Halley thought that once he had mapped those lines, they would be a long-term, invaluable record of the angle of deviation at sea between geographical north and magnetic north. In fact, they were out of date almost as soon as they were published because the field is so changeable.

But today, even with the three centuries of understanding and information that have followed Halley's work, geophysicists can only predict the movement of the field lines at the surface for as much as five years. Not more. Anything past that is soothsaying. So, every five years, geophysicists gather as a community to work out their mathematical forecast for the next half decade and publish it, making it freely available to all. Called the International Geomagnetic Reference Field, each version is out of date by the time the new one is made. The models are extraordinarily detailed. Finlay was lead author of the forecast in 2010. Knowledge of the magnetic field's direction is an essential part of many modern navigation and orientation applications, and for industries working underground. It works for aviation during those times when GPS satellite systems don't work well enough or fail altogether. Even smartphones rely on the magnetic models Finlay and his colleagues produce.

And this five-year forecast is for the field as a whole on the surface of the Earth. Predicting what will happen in the dynamo or to the gyre within the core is a much more remote frontier. Trying to predict what the poles will do, even tougher. And here's the sobering truth: While geophysicists are avidly examining past reversals for clues about the next one, there's also the possibility that every reversal acts differently, or that the dynamo itself is changing and that the clues from the past won't solve the riddle.

CHAPTER 23
the worst physics movie ever

At the Nantes meeting, the question of whether the poles are in the throes of a reversal was like Banquo's ghost: unwelcome and mainly invisible. There was no session on it. Several posters touched on it but only at a slant, as if in code. When, just once, a scientist asked a direct question about it after a talk, the presenter neatly deflected. Yet everyone there knew that the responsibility to find out—if that were possible—rested uniquely with this community of scientific minds and with no one else. And not just the responsibility to find out but also to warn society if the switch were to show signs of arriving, just as climate scientists have had to shoulder the task of explaining the effects of high carbon dioxide concentrations in the atmosphere. Instead, it felt as if the geophysicists were chastened. Having seriously considered the possibility that a reversal was looming in the heady decades after satellite data began streaming in, they had pulled back.

I sought out Peter Olson, an emeritus professor at Johns Hopkins University, in Baltimore, Maryland. One of the world's most eminent geophysicists, he wrote a famous commentary for *Nature* in 2002 titled "The Disappearing Dipole." In it, he referred to the "amazingly rapid decline" in the dipole's strength in the past 150 years. While carefully noting that it was "premature" to suppose that the dipole would continue to decline until it was gone, he pointed to the growing reversed-flux patches and wrote that they suggest an attempt at a reversal "may be underway." That article accompanied another famous one in the same issue of the journal, also suggesting that the satellite data show the dynamo might be gearing up for a reversal. In *Nature* in 2008, Gubbins added that our current situation "could be the start of a reversal, but we have not yet reached the point of no return." None of this was declarative—or too much different from what Finlay was able to conclude when I saw him—but it was a little bolder than the dominant thread today. And the scientific idea that a reversal was imminent was sexy enough to have penetrated the public discourse. There was a spate of articles in the popular press in the 1990s and even a 2003 movie, *The Core*, starring Aaron Eckhart and Hilary Swank. Its premise was that the core stops rotating, the Earth's magnetic shield fails, and deadly microwave radiation immediately starts killing people by halting their pacemakers. Civilization is also threatened. It's been panned as the worst physics movie ever.

Olson looked slightly resigned when I asked him about a reversal. We were assembling in the expansive main floor of the conference center in Nantes with the other participants, to have a group photo taken. It was hot and some people wore shorts, sandals, and short-sleeved shirts. Sure, this could be the foreshadowing of a reversal, Olson said. But it's more likely just a phase. If the South Atlantic Anomaly were to expand to 30 percent of the Earth's surface, then that's a reversal. (Today, depending on how you measure it, it is just over 20 percent, up from 13 percent in 1955.) Will that happen? It's simply not predictable, he said. It's like predicting when and where a hurricane will hit a decade from now. One of his areas of research has been on the effects of a reversal on mass extinction. He hasn't been able to find a link. "It's not as if people haven't thought deeply about this," he said, adding that it would make the whole conference—here he gestured to the throngs of scientists—more relevant. As I thought about it, I realized that the meeting did not have the air of urgency one finds at, say, a conference on climate change, where the effects of the scientific phenomenon are playing out in every backyard in the world, sometimes with deadly consequences.

Next, I tracked down Cathy Constable, a geophysicist at Scripps Institution of Oceanography at the University of California, San Diego. An international leader in the art of using statistical techniques in geophysics, she and her co-investigators have been tireless in constructing models of the magnetic field that go back in time millions of years. She's used them to examine what the field looked like during previous reversals and has compared those scenarios to today's. During a break between sessions, I asked her if we are in the throes of a reversal. "Hell no!" she said. "Not in my lifetime!"

In 2006, Constable and Monika Korte from GFZ German Research Center for Geosciences in Potsdam published a paper meticulously outlining the case for an imminent reversal and doing the maths to assess the likelihood. It was like reading a crisp legal brief. The assumption was that the more we know about past reversals, the better we can predict whether one is happening now. So, is a reversal overdue? The last one was 780,000 years ago and reversals have been happening about three times every million years for the past 90 million years. Therefore, some say, it's time. Constable and Korte crunched the numbers and found that contention statistically suspect. Maybe. Maybe not. Over time, an interval of more than 780,000 years between reversals is not wholly unusual.

What about the argument that the dipole is decaying fast? Constable and Korte pointed to the fact that while, yes, it is waning, its pace while doing so is in line with what's been going on for the past 7,000 years. Nothing unusual there. At other times during the past 7,000 years, it has decayed just as fast or even faster without prompting the poles to flip. Not only that, but it is still strong compared to the dipole's intensity at the moments of other reversals. It's even strong compared to the long-term average over the past 160 million years; it's nearly twice that figure. The role of the South Atlantic Anomaly in triggering a reversal was a little tougher to analyze. Their best conclusion: There is an absence of evidence to support a looming reversal.

But the fact that so far there is no certainty on the timing of a reversal doesn't stop geophysicists from trying to figure it out. For example, the north magnetic pole, which has always wandered, has begun moving at a full gallop to the north-northwest at the pace of about 55 kilometers a year. (The south magnetic pole, by contrast, continues to meander sedately.) An animated video of its track just since 1999 shows a breathtaking race across the High Arctic. It's a clear indication that the field is changing rapidly, that something is afoot in the outer core.

Not only that, but a new batch of papers has used novel methods to come to different conclusions about a reversal. A French study looked at the past 75,000 years of sedimentary and volcanic data as well as ice cores from Greenland, where radioactive isotopes of beryllium and chlorine had been deposited over time. The concentration of those isotopes, created when cosmic rays battered the Earth's upper atmosphere, is a good measure of the intensity of the Earth's dipole. The more there are, the lower the dipole's strength. The authors, one of whom, Carlo Laj, traveled to Pont Farin in 2002 to redo Brunhes's measurements, found excellent matches in the records for the Laschamp near-reversal, a nod to the precision of the method. Their conclusion: The field is decaying so fast that a reversal may be irreversibly under way, but the poles themselves won't reverse for at least five hundred years. But since the risks of a reversal lie not in the actual shift of the poles but in the strength of the magnetic shield to protect the Earth from dangerous radiation, this finding implies that the next five hundred years or more could be the ones to watch. Not much comfort.

A study by two Italian researchers took an altogether different, highly controversial approach. They used a theoretical systemics approach to examine the geomagnetic field, looking at how it interacts with other systems governing the planet. The behavior of the South Atlantic Anomaly in particular led them to conclude that today's field is "rather special" and is approaching a critical transition. They even put a date on the point of no return: 2034, give or take three years. That's not when the reversal will happen, but when it becomes inevitable.

At the University of Rochester geophysicist John Tarduno and others did an ingenious study examining the burned clay huts of Iron Age Bantu-speakers who lived in villages along Africa's Limpopo River beginning in about 1000 CE. These are some of the few ancient readings from the southern hemisphere. When the rains failed, these early farmers torched their storage huts in a cleansing ceremony, heating up the magnetite-rich clays well past their Curie point. As they cooled, they took on the magnetic memory of the day. Tarduno, working with archeologists, has discovered that this part of Africa displayed a low field seven hundred years ago. The field then regrew in strength before weakening again to form part of the South Atlantic Anomaly.

The findings are significant, Tarduno argued, because the weak patch in the magnetic field, then and now, lies over the edge of an unusual formation in the mantle at the boundary of the core. It's a place millions of years old with steep sides, where seismic waves move with unusually low velocity. Tarduno posited that this piece of mantle affects the movement of molten iron in the outer core, changing its magnetic flow. In turn, this shifts the field's direction, producing the reversed-flux patches now visible, and saps the field of strength on the surface. Tarduno's proposal was that rather than being triggered by random phenomena in the core—or related to Finlay's gyre—reversals might be triggered by this oddity in the mantle, particularly if several reversed-flux patches link up. While Tarduno stopped short of saying a reversal is nigh, he stressed the dramatic decay of the dipole over the past 160 years, calling it "alarming."

And new findings keep emerging, questioning the basic understanding of how the Earth's field works. A fascinating paper published in 2017 examined the handles of clay jars made in the Levant near Jerusalem between 750 BCE and 150 BCE. The handles were stamped with royal Judean seals when they were still wet, which means the date of their firing can be pinpointed with a high degree of precision and therefore so can the date of the magnetic field they record. This degree of exactitude in the rock record is rare. Just before 700 BCE, the field's intensity spiked up by about 50 percent. At that time, the field was already strong compared to today, so with the spike it became nearly twice as strong as it is now. The odd thing is that it decayed abruptly too, waning by more than 25 percent in just thirty years. That's far more rapid change than geophysicists have believed the outer core is capable of. If it's real, it suggests a degree of volatility previously unimagined.

For their part, French geophysicists Jean-Pierre Valet and Alexandre Fournier adjured their colleagues to keep heart. In an exhaustive review paper, they argued that the answer to understanding what happens during a reversal lies in closer examination of sedimentary rocks. They argued for better techniques in studying rocks' magnetic memory, especially to track the field in the throes of a transition. Perhaps greater use of new magnetometers that test rock samples so small as to be nearly microscopic. Perhaps the beryllium isotope readings. "Despite many unresolved questions we are far from pessimistic and consider the quest for a proper description of polarity transitions to not be hopeless," they wrote.

Back and forth. Back and forth. Like so much of the lengthy

investigation of the Earth's fickle magnet, the questions outstrip the techniques that could provide definitive answers.

So where to go from here? Finlay had his eyes trained on what might emerge from a whole new angle of investigation. Not squeezing more information from more ancient rocks. Not building ever more detailed analysis of what rocks are saying. Not calculating ever more realistic numerical models of the field. Instead, trying to physically reproduce the self-sustaining dynamo in the heart of the Earth. That means rather than being stuck on the surface, or at the boundary between the mantle and the core (which is as deep as the maths will allow maps to be constructed), geophysicists could go even deeper, right into the mysteries of the outer core itself. One of the promising experiments is in a lab in Maryland run by Daniel Lathrop. The whole geophysical world, Finlay confided, is holding its breath waiting to see if Lathrop's experiment will "dynamo"— that is, make a dynamo on its own. Perhaps a plot for *The Core II*, starring Hilary Swank?

CHAPTER 24

the great hazardous spinning
sphere of sodium

Torrential rain, thunder, lightning, and hail had pummeled the University of Maryland in College Park the evening before I was to visit Daniel Lathrop's lab. The expansive greens of the Georgian-style campus were sodden the next morning. The air was heavy. But the leaves of the chestnut trees were open and full of promise. Lathrop entered his office at a canter, skidded to a stop, and immediately started talking fast. Tall and rangy, dressed in khakis, he seemed as kinetic and nonlinear as his subject. To wit: the machine in the core of the Earth that continually creates and destroys the planet's magnetic field, also known as the geodynamo.

His work began with a conundrum. The Earth's core is not permanently magnetized, he explained, plopping down in a chair, a well-used espresso machine to his right and a backpack tossed onto the floor at the other end of the room. He was lining up a trip to take his whole family to Calgary, Alberta, in the Rocky Mountains, on summer vacation, where he planned to rent a big RV and see as many of Canada's national parks as humanly possible in two weeks. The core can't be permanently magnetized, he continued, because its temperature is far past the Curie point. Yet the Earth has a magnetic field. So where does it come from? And how do you understand the magnetic core well enough to describe it mathematically, bring the maths to life in the lab, and then make predictions about its behavior that you can apply to the real world?

The wild card is its turbulence, he told me. Maths has long been able to describe how a liquid flows in an enclosed space. Sometimes it's calm; sometimes more agitated. That agitation is called turbulence. The more turbulent something is, the harder to predict, the more nonlinear. Lathrop, ever the geophysics professor, pointed to the Earth's atmosphere. That huge storm we had last night with the hail and lightning? That was turbulence in the fluid medium of the atmosphere. But in the air, you can see what's going on. And if you create a model to predict next week's weather and you're wrong, you can adjust the model to reflect what happened. That makes the model more and more accurate over time. In the core, it's much tougher to see the storms in the first place and therefore it's tougher to create a good model for predictions. And it takes longer to find out whether you're right. It could take tens to thousands of years, which doesn't exactly fit into a scientist's career plans, Lathrop noted wryly. Plus, the core is far, far bigger than the atmosphere and therefore has a lot more turbulence, making it even harder to forecast.

Why is the fluid in the outer core turbulent? What purpose does it serve?

He leapt up and wrote a simple equation on the board beside his desk. ("This is the only equation I'll write for you," he promised, chuckling.) It's a formula for what's known as the Reynolds number, which predicts how fluids will flow using the variables of velocity times size divided by viscosity. It shows that anything really big has a nonlinear flow. Not only is it turbulent, but it *must* be turbulent. It's the way nature works. More important, it's how physics works. For example, blood flowing through capillaries has a small Reynolds number. It flows in a relatively calm, predictable way, with little turbulence, because capillaries are small. Clouds have a high Reynolds number and therefore their flow is nonlinear, sometimes resulting in storms or hurricanes. But the flow in the Earth's core is almost incomprehensibly nonlinear because the core is so large.

"You've got to expect it's gonna have weather," Lathrop deadpanned. In fact, the equation to get the Reynolds number for the core can be written, but not solved, although scientists don't like to admit that, he said. What that means is that there is at present no scientific way—either theoretical or mathematical—to predict the future of the Earth's magnetic field beyond the five years that Finlay and his colleagues can calculate for the International Geomagnetic Reference Field. It's like the weather. Forecasters can give us a pretty good idea of what the weather will be like tomorrow and next week and even two weeks from now. But ask what the temperature will be like on New Year's Day in a decade, and meteorologists resort to generalities.

And that brought Lathrop to his current experiment, the latest in a string in which he has tried to replicate the Earth's dynamo. Joseph Larmor, an Irish mathematician, wrote a two-page paper in 1919 suggesting that both the Earth and the sun might have a self-sustaining fluid moving inside them. This was before Harold Jeffreys had discovered that the core was fluid and before Inge Lehmann had found the inner core. But Larmor used Michael Faraday's experiments in the basement of the Royal Institution as his leaping-off point to create the vision of the Earth as an electrical generator. The Earth's interior was shedding heat through convection into rotating molten metal whose atoms had unpaired spinning electrons. The convection of heat produced a system of electrical currents flowing in the liquid, which, as Faraday had shown, produced a magnetic field. Larmor's idea was hotly contested by "anti-dynamo" researchers and mainly ignored until after the Second World War. A series of brilliant numerical models using some of the world's first supercomputers finally produced a full-scale simulation of the geodynamo in 1995. Several times, the dynamo's field spontaneously reversed direction. This model by geophysicists Paul Roberts of UCLA and Gary Glatzmaier, now at the University of California at Santa Cruz, showed that the outer core often tried to trigger reversals but that the inner core usually blocked them. That suggested the enigmatic inner core held the key to reversals. To Lathrop, the next step was to see if he could create a real-life dynamo in a lab.

It was set up in a neighboring building on the campus and as we trotted there, he explained how it worked. Lathrop has spent a lot of time with journalists. In fact, he was dashing into a second hour-long interview right after he finished with me. He's perfected the art of explaining what he's doing in the lab without agonizing about what he's finding. To him, science is an endlessly fascinating exercise in slaking curiosity, endpoint uncertain. "I try not to have very strong personal desires about what the science shows us, because that could lead to bias," he said. In fact, he is so nonchalant about outcomes that he's fond of deconstructing the whole idea of scientific certainty: "All science is provisional," he told me, shrugging.

It took him eight years to work out the details of this latest experiment: a stainless-steel sphere three meters in diameter containing a hollow inner sphere one meter across, roughly the proportion of the Earth's inner core to its outer core. Each sphere could rotate independently and was hooked up to a motor. The outer sphere was bound with magnetic coils. The space between the two spheres was filled with 12.5 tons of sodium. Sodium is a silvery-white metal so soft you could cut it with a knife. It has one unpaired electron in its outermost filled orbital. Sodium is one of several elements that Humphry Davy discovered in the early 1800s as he experimented with his voltaic piles and the then new process of electrolysis, which uses electrical current to tear molecules apart. It is the best liquid conductor of electricity on Earth, a proxy for the molten iron and nickel in the outer core.

It is also lethally explosive, including at room temperature. Any water touching the sodium, even a drop of sweat, can cause it to react. At higher temperatures, it can combust on its own, producing sodium peroxide smoke caustic enough to burn skin and damage lungs. Sodium is used to cool nuclear reactors and its unusually high volatility has led to an extensive history of serious sodium fires in those reactors.

There was so much sodium in Lathrop's sphere that it took his team a day and a half to get it above its melting point of 98 degrees Celsius—nearly the boiling point of water—before they could run the experiment. They started the melt on a Monday morning each month and began to spin the spheres by Tuesday afternoon, letting them run until the end of the day on Friday. After that, they spent three weeks crunching the data and tweaking the experiment. The spheres were enclosed in a huge metal box, centered in a cavernous laboratory space. Stairs alongside the sphere reached a platform on the top, where lab assistants had set up a computer. When the sphere was twirling, no visitors were allowed in the lab and the team members were in a safety control room a few meters away with their computer terminals. As we walked in, I asked: "Is this dangerous?" Lathrop replied: "I prefer 'hazardous.'"

The question behind running the experiment—apart from Lathrop's stated determination to have "no fires and no fatalities"—was whether one could make a self-sustaining dynamo as similar as possible to the Earth's within the liquid sodium in the sphere. And then see how it behaves. How is the turbulence shaped by the rotation? How does turbulence affect the sodium's ability to conduct electricity, if at all? Over the longer term, if the sodium "dynamos" on its own, Lathrop and his team may be able to witness a reversal within it. They may even be able to figure out how to predict what the field will do. So the team spins the spheres fast to drive turbulence in the sodium, a proxy for the spin of the Earth. At the same time, the team imposes a small magnetic field onto the spheres, like sowing seeds in a furrow, to see if the sodium will produce its own larger, self-sustaining magnetic field. So far, the flow of the sodium is able to amplify the imposed magnetic field by a factor of ten. But so far, there's no self-sustaining dynamo and no reversal.

So, no dynamo, no solution to the Reynolds number to describe the turbulence of the core, no ability to predict what the magnetic field will do, no way to say whether the field is in the process of re-versing or, if it is, when it will happen. In fact, no precise idea of what the field looked like during past reversals and no certainty that the dynamo is operating the same way now as it has over the past bil-lions of years it has existed.

And without predictions, Lathrop said, we can't prepare. Maybe we don't need to prepare, he mused. But it would be good to know for sure.

Prepare for what? Between Lathrop, Finlay, Constable, and oth-ers, there's quite a list. The concern is what happens as the field is in the process of reversing. That's when the field protecting the Earth dies down to perhaps only one-tenth of its normal strength. The magnetosphere, that stretchy web of invisible lines surrounding the Earth that makes our planet a galactic sanctuary from radiation, could show up in a more complicated pattern, Lathrop said. What will it look like when the dipole is beaten back so far that other mag-netic poles are present? What will its protective force look like then?

While Constable scoffed at the idea that a reversal is imminently on the way, she also said she is far from sanguine. The field is demon-strably not stable. She pointed to the paleomagnetic evidence that we have been living in an unusually strong magnetic field for the past few hundred years, exactly at the same time as we have developed electromagnetic systems of technology and become dependent on them. When the field weakens and solar radiation penetrates closer to the surface of the Earth, those systems could be vulnerable to attack. That's even without a reversal. When she looks back into the archive of the past seven thousand years, she sees fluctuations that have been large enough to have serious repercussions for society.

As for Finlay, he said he didn't lie awake at night tossing and turning about fallout from the decaying dipole. In fact, he abhorred the alarmism that sometimes accompanied media discussions of a flip of the poles. His best analysis was that a reversal would be a leisurely process taking many hundreds or a few thousand years. Like Constable, his most immediate concern is how a weakened field will affect technology. He pointed out that the last time the field reversed, an advanced society based on electromagnetic systems wasn't around. If the field continues to weaken, society is going to have to think about how to modify technology to protect it from surges of solar radiation. When? He gave a scientist's answer: We would be wise to start preparing as soon as possible.

PART IV

switch

[A scientist] has to have the imagination to
think of something that has never been seen
before, never been heard of before. At the same
time the thoughts are restricted in a strait
jacket, so to speak, limited by the conditions
that come from our knowledge of the way
nature really is.

—Richard Feynman, *Lectures on Physics,* early 1960s

CHAPTER 25

looking up

Boulder, Colorado, lies at the seam between the prairies and the Rocky Mountains, where the clouds, if there are any, cast sharp shadows on the land. The day I traveled there to meet Daniel N. Baker, it was hot and the flattened faces of the mountains—called flatirons, as if they could be cajoled into pressing giant shirts—stood stark against the sky. I could still see a faint sliver of moon at midday. Underfoot, as I navigated the sprawling campus of the University of Colorado at Boulder, the smell of low-slung hedge cedar rose up, mixed with the fleeting scent of spring's first lilacs. It had been fourteen months since I had wandered the ancient streets of Clermont-Ferrand with Jacques Kornprobst, and I was nearing the end of my research.

Like Brunhes with his fin-de-siècle observatory that reached up from the Puy de Dôme volcano, the founders of the High Altitude Observatory in Boulder were drawn to the peaks. As the space age roared ahead in the decades after the Second World War, the Boulder observatory drew academia, research, and industry together in this picturesque mountain enclave of a hundred thousand, building what had been a scientific backwater into an international power base for parsing how outer space affects humanity.

Baker was one of the scientists who came here for what the mountains and the sky could tell him. The stars. The planets. A desire to understand an eclipse of the sun led him, as an undergraduate student in the 1960s, to the University of Iowa to study with James Van Allen. In 1958 Van Allen, then the world's most famous astrophysicist, had discovered banks of radiation trapped by the Earth's magnetic field in two fat crescents straddling the Earth's magnetic equator. They're known as the Van Allen belts and are highly stable, reliably storing radiation for long periods so that it isn't let loose closer to our planet's surface. Most of the radiation is charged electrons and protons created by cosmic rays crashing into atoms in the upper atmosphere and tearing them apart. The discovery of the belts marked the beginning of the space age. It founded the formal scientific study of the Earth's magnetosphere, which is the space component of its magnetic field. It also put Van Allen on the cover of *Time* magazine. Twice.

In his sophomore year, Baker took a course in modern physics from Van Allen. Van Allen offered him a job. Baker has been involved in space exploration ever since, often as a collaborator with his mentor. In 1994, after a stint at NASA, Baker became the director of the Laboratory for Atmospheric and Space Physics (LASP) at UC Boulder. His curriculum vitae runs to more than 130 pages and, by 2017, included more than nine hundred published papers—a towering scholastic output sustained over more than four decades. The odd thing about it to a science journalist is that Baker's writing style is so plainspoken, even chatty, that you can pick up one of these papers and unmistakably hear his voice in it. Most scientific writing is far more anonymous.

Baker lit up when he talked about Van Allen, who died in 2006, just short of his ninety-second birthday. So famous, he told me, yet so disarming. Reporters used to mob him wherever he went. Everybody wanted to talk with the great Dr. Van Allen. They would ask him: What are the Van Allen belts good for? He would answer: To hold up Van Allen's pants! When Van Allen turned ninety, his family and friends threw him a big party back in Iowa City and invited Baker, the star student, to give an after-dinner address to the hundreds gathered. Baker put together a funny faux-Letterman top-ten list. A few weeks later he got a stern letter from Van Allen. Ever the teacher, he quibbled with some of Baker's analysis, he told me, smiling fondly.

Baker's focus is how the sun's radiation affects the Earth. The two bodies have a complicated, interlocked relationship, mediated by the elastic magnetic fields of each. By radiation, he doesn't mean the long, slow, benign waves of heat, light, and color, but the dangerous, invisible facets of the electromagnetic field: the very short, high-frequency waves with enough energy to harm an atom or cell, such as X-rays and gamma rays. "Part of my job is to make the invisible visible," he confided. "I see things in wavelengths most people are not sensitive to."

This is called ionizing radiation because it is powerful and swift enough to knock electrons out of their orbitals, creating ions, a term Faraday coined. (An ion is an atom or molecule that has a different number of electrons than protons, making it electrically charged. That lack of electrical balance makes it eager to restore its own balance, meaning it is apt to interact with another atom or molecule.) This stuff is bad for life on Earth.

Baker is particularly interested in a different form of the sun's energy that is not part of the electromagnetic field: highly damaging solar energetic particles. The sun is too hot for atoms to survive, even as gas, so it is mainly made of plasma, the fourth and hottest state of matter, after gases, liquids, and solids. The sun's plasma is essentially the components of hydrogen and helium atoms. It's electrically charged moving particles: protons, electrons, and ionized nuclei. As they move, they make electric currents and therefore a magnetic field. The sun's plasma is so hot that some of its most ener-getic particles routinely break free from the gravitational pull of the sun and fly into space. That's called solar wind, and it's the force that crushes up against the side of the Earth's magnetosphere facing the sun. Occasionally, high-speed streams of solar plasma break through holes in the sun's outer atmosphere, or corona, and pummel the Earth for hours or days.

The sun's interior is restless and volatile. Its continual contortions stretch its magnetic field until, suddenly, the field lines snap back into place, releasing enormous amounts of magnetic energy. Sometimes, that shows up as a flare on the corona, a burst of electromagnetic waves of all lengths, including the longer radio waves, which can last from several seconds to several hours. Some of those waves are visi-ble as white shapes during a solar flare. The waves travel to the Earth's upper atmosphere in eight minutes and, if they're strong enough, can disrupt radio transmissions.

But solar magnetic disturbances are also capable of producing ex-plosions of part of the corona: violent, targeted, billion-tonne masses of magnetized plasma moving at 3,000 kilometers a second or more toward the Earth. They call them coronal mass ejections. NASA has produced images of these ejections, and they look like the hot red flame from a dragon's maw roaring toward an Earth the size of a flea. Coronal mass ejections can also produce shock waves. When the magnetic field of the ejection runs opposite to the direction of the Earth's, it can trigger storms in the Earth's field. One manifestation of these geomagnetic storms is auroras, like the curtains of green light I saw pulsating in the night sky above King William Island in the Arctic. Solar flares and the shock waves from coronal mass ejections can also produce solar energetic particles. That's even faster and far more damaging radiation.

Like solar plasma, solar energetic particle radiation is made up of protons, electrons, and high-energy nuclei. They are electrically charged and moving, which means they make electrical currents and therefore magnetic fields. They are also ionizing. Ionizing radiation—whether solar energetic particles or electromagnetic—is like an invisible bullet of energy. It can obliterate whole swaths of DNA as it passes through tissue. It can cause cancer, genetic defects, radiation sickness, and death. And in addition to episodes of solar radiation the Earth is continually attacked by galactic cosmic rays, which come from outside our own solar system, possibly from the explosion of supernovas in our galaxy, the Milky Way.

Our protection from all this damaging ionizing radiation comes from the magnetosphere, the atmosphere, and the two Van Allen belts. And those, in turn, depend on the dynamo in the core of our planet. Early in its life, our sister planet, Mars, had an ancient internal dynamo churning out a protective magnetic shield that also episodically switched direction. That shield allowed it to have a thick atmosphere and bodies of water on its surface. By about 4 billion years ago, the dynamo died. A leading theory is that its metallic core cooled down so much that the all-important heat convection needed to make electrical currents stopped. Or the culprit could be that its tectonic plates fused—assuming it had them—making the "stagnant lid" crust the planet has now. That would mean that Mars wasn't able to shed heat effectively from the core. Again, convection would cease. The third hypothesis is simply that the dynamo ran its course after the core shed enough heat, solidified, and resulted in an outer molten core too insubstantial to sustain electrical currents. And while the reason for the dynamo's death is still being investigated, NASA's recent MAVEN (Mars Atmosphere and Volatile EvolutioN) mission to Mars, which Baker was involved in, confirms the upshot: As the dynamo waned, the sun's ferocious stream of wind and ultraviolet radiation scoured away Mars's atmosphere. Without an atmosphere, including enough carbon dioxide gas, the planet isn't expected to support life. It is now too cold and too vulnerable to all that radiation, but missions continue to search for evidence of it.

NASA's Juno mission to Jupiter, launched in 2011, which Baker is also involved in, is investigating that planet's powerful magnetic field and radiation belts for more clues about how all the solar system's dynamos work. Close-ups, relayed in 2017, are haunting. Looking at Jupiter's south pole is like looking into the core of a swirled blue marble, gigantic cyclones peppering its inner reaches. More of Lathrop's turbulence, but on a far vaster scale. Fascinating, but humbling too, to catch a glimpse of such a powerful magnetic field. Jupiter's is about ten times as strong as Earth's, driven by a dynamo likely nestled in a metallic hydrogen core. All this new information is helping dynamo theorists, who once only had Earth's field to examine, realize how complex any dynamo is, Baker said.

But to Baker, looking up also means looking forward. He has been on a long-standing quest to better predict solar storms, urging society to learn how to protect itself against these events. That led him to train his attention on the game plan of the Earth's magnetic poles. What will happen when the poles switch places and the magnetosphere ebbs, along with the magnetic field? The very structure of the Van Allen belts, which depends on the terrestrial dipole, will be deformed. During a reversal, the belts are expected to become more complicated banded structures, far less stable, less well defined, and much less efficient at trapping radiation.

In addition, the Earth's field itself will not be strong enough to defend us from as much radiation, in the form of solar wind, solar flares, coronal mass ejections, solar energetic particles, and galactic cosmic rays. And the lack of protection would likely be at least a centuries-long, global phenomenon. Maybe millennia-long. Not an earthquake or tsunami or volcano that comes, wreaks destruction, and then leaves people to heal and rebuild. It would inflict damage across generations. To a scientist who has spent a lifetime immersed in the precise and peculiar possibilities of harm from cosmic radiation, this picture, which has been sharpening over the past couple of decades, presented a very large red flag.

Baker and I had spoken at length by phone about the possible reversal of the poles, and I was in Boulder to meet with him in person. An early riser, Baker was often at his desk hours before any of his five hundred employees parked in the laboratory's expansive parking lots. Not only is he the head of LASP, but he is a faculty member in both the physics and astrophysical and planetary sciences departments. When I met with him, LASP, which designs, builds, and tests space hardware for space missions, was operating four missions for NASA. His organization has between fifty and sixty grad students at any moment. I had been prepared to encounter a tightly scheduled chief executive. He was the opposite. He wanted to talk. He was worried.

He ticked off the reasons. A space physicist, he had been monitoring the growth of the South Atlantic Anomaly, just as geophysicists had, but using data from NASA's *SAMPEX* (Solar Anomalous and Magnetospheric Particle EXplorer) spacecraft. Rather than measuring the magnetic field itself, it measured concentrations of charged and energetic particles several hundred kilometers above the Earth's surface—in other words, what the depression in the field let in. The spacecraft found not only that the anomaly has moved over the course of twenty years—it lies squarely over Brazil at the moment—but also that it is growing and that the field above it is weakening, a match for the calculations geophysicists are making with the Swarm satellites. Not only that, but the magnetic north pole is moving fast and the field as a whole is weakening. The kicker for Baker was that you can measure the change within the twenty-year lifetime of a spacecraft, not in the leisurely time frame of many thousands of years that the geological record normally trades in.

So, unlike so many of the geophysicists who are carefully assessing whether the poles will flip, Baker leapt ahead to what he called "plausible scenarios." If this is the beginning of a reversal of the poles—or even if it might be—he was compelled to consider the implications. It was his training. What would our world look like right now if the magnetic field dwindled to just one-tenth of its strength?

As the late physicist Richard Feynman said, scientists are bound by what is known. Humans were not around during the last reversal 780,000 years ago. There are no written or oral records to consult. Nevertheless, the past has left a few clues about what has happened during previous reversals. And scientists have seeded indications in a few other studies about what could happen, if you extrapolate. By poring through the scientific evidence for these clues, looking at the evidence of how solar storms already affect our world, and then applying those lessons to the future, we can catch a glimpse of what might come. It's like following a trail of breadcrumbs.

Two questions stand out. What do we know about how more intense solar weather will affect civilization? And what do we know about how it will harm living creatures, including humans?

horrors the lights foretold

The idea that damaging radiation from space can affect the Earth is not purely academic. Bursts of radiation occasionally pierce the Earth's magnetic shield and atmosphere even now during solar storms. These unpredictable storms tend to coincide with perturbations in the sun's own magnetic field, which peak and wane over the eleven years or so between its pole reversals. Sometimes they appear more randomly. To Daniel Baker, they provide the first pointers to what an unprotected future might look like.

Occasionally, when the events are powerful enough, damaging radiation scatters through the atmosphere and strikes the surface of the Earth. These events are known as "ground level enhancements" of cosmic rays, one of the scientific euphemisms that pepper the literature on this topic. Since 1950 more than seventy ground level enhancements have happened, based on an international network of radiation monitors on Earth, akin to solar Geiger counters. A dominant concern is how these events affect airplane passengers and crews—particularly pregnant ones—as well as the equipment needed to fly the aircraft.

But in addition to how solar storms affect avionics and living tissue is the way they can damage hard-to-replace electrical infrastructure, pipelines, and other industrial technologies, particularly when their systems are interconnected. Those interdependencies are growing day by day, as societies link electrical systems ever more tightly together, unaware of the risks.

Two relatively recent solar storms have been minutely examined for their lessons on how vulnerable modern technological systems are to solar radiation. The first, on March 13 and 14, 1989, knocked out power in Quebec when storm-induced currents overwhelmed one part of the system and its built-in protective mechanisms kicked in, shutting down first one section of the grid and then every other in succession. Six million people were without power for nine hours. A nuclear unit transformer in Salem, New Jersey, overheated and had to be taken out of service temporarily. In the UK, the same storm damaged two electrical grid transformers. Scientists call it the day the sun made things dark.

A broader event, called the Halloween magnetic storm of 2003, happened after seventeen flares erupted in neighboring areas of the sun, many of which were followed by storms of radiation. A large one exploded on October 28, quickly followed by a coronal mass ejection. That mass of plasma was clocked at 2,000 kilometers a second by a nearby instrument, which functioned until protons from the flare blinded it. The next day, another huge flare burst forth, followed by yet another wild coronal mass ejection. Once their energy hit the Earth on Halloween, October 31, electrical engineers across North America raced to protect the electrical grid by shutting down key parts of it. Transformers in South Africa were badly damaged, and one in Malmö, Sweden, shut down, leaving fifty thousand people without light for an hour. At least thirteen nuclear power reactors took measures to protect their equipment from damage.

The event spawned the first-ever radiation alert to aircraft from the Federal Aviation Administration. Spikes of radiation entered the Earth's atmosphere at the poles, where the magnetic field lines exit and enter. Flights over polar routes were redirected because passengers and pilots were at risk. Also, airlines could not rely on a key global positioning system for precision landing because satellites orbiting the Earth had shut down. Geomagnetic measurements for oil and gas drilling failed. Magnetic and geophysical surveys ceased to function. The US military had to cancel a mission at sea because the satellites that governed their communications were disabled. The $640-million Japanese scientific satellite ADEOS II was lost to space, carrying a NASA instrument worth $154 million. Astronauts at the International Space Station four hundred kilometers above the Earth (inside the Van Allen belts) had to duck for cover and hope for the best under layers of extra anti-radiation shielding that the Russians had thoughtfully provided. That action cut the radiation the astronauts experienced by about half.

The energy from the Halloween event pulsed through the galaxy for more than a year. The Mars *Odyssey* spacecraft was hit with so much radiation from the event's solar energetic particles that its sensors shut down. The solar flare was so strong it outshone the stars, and the Mars spacecraft couldn't navigate using them as reference points. Probes near Jupiter and Saturn recorded the event and so, much later, did the *Voyager 2* spacecraft, 11 billion kilometers from the sun.

Far more serious—and therefore more telling because they are more akin to what the Earth will experience when the poles are reversing—are the biggest of the superstorms. They are far more intense pulses of energy tossed off from the sun that temporarily but severely weaken the Earth's magnetic field. Scientists are aware of only two of this type of superstorm since they began chronicling the sun's activity hundreds of years ago. The first is the Carrington event, named after Richard Carrington, the astonished British astronomer who watched it unfold. Baker has studied the event, which happened in 1859, the same year Boulder was founded and that Darwin published *On the Origin of Species*.

That was only twenty-eight years after Faraday had made his induction ring in the basement of the Royal Institution in London. Even when Faraday's experiment was successful in producing electricity from a magnet, the idea of harnessing that electricity to power society was inconceivable. But by 1859, the first transatlantic electric cable had begun transmitting telegraph signals from Europe to North America. More than 100,000 miles of telegraph lines connected stations across those two continents and in Australia. It was the first continental-scale electric technology humans had created.

The event started with sunspots on the face of the sun on August 28. On September 1, Carrington saw "a singular outbreak of light which lasted about 5 minutes." It was the first solar flare reported by a human. Carrington even drew pictures of it, small wormlike figures. The flare was followed by what we now know was a powerful coronal mass ejection—a blast of electrically conductive, magnetized plasma—and then by a storm of solar energetic particles. Slower than the electromagnetic waves of light that marked the solar flare, the plasma took seventeen hours and forty minutes to reach the Earth. We have learned, based on ice-core analysis, that it was the most dangerous solar radiation event of the previous five hundred years. It is considered the benchmark for worst-case, life-threatening exposure to solar radiation for astronauts. Not only was the storm savagely strong, but its magnetic field also ran opposite to the Earth's. This was the recipe for an acute magnetic storm on Earth.

Stupendous auroras lit up the skies during the dawns of August 29 and September 1 and 2. Rather than in the usual narrow oval bands around the two poles, dramatic auroral displays could be seen in countries just a few degrees from the equator in both the northern and southern hemispheres, as well as in other parts of the world where they were unfamiliar. "The light appeared in streams, sometimes of a pure milky whiteness and sometimes of a light crimson. . . . The crown above, indeed, seemed like a throne of silver, purple and crimson, hung or spread out with curtains of dazzling beauty," wrote a Washington reporter. "The whole sky appeared mottled red, the arrows of fire shooting up from the north like a terrible bombardment," wrote a correspondent from Ohio. People were terrified. They believed their towns were on fire, and some raced to houses of worship to pray for the prevention of whatever horrors the lights foretold.

The new telegraph system, with its electrical lines, became a target for rampaging currents produced by the magnetic disturbance. Lines went down in New York, Boston, Philadelphia, Washington, Massachusetts, London, Brussels, Berlin, Mumbai, throughout Australia, and in every single telegraph office in France. In Pittsburgh, batteries connected to the telegraph wires emitted streams of fire. In Sweden, they sprayed electric sparks. In Norway, they sparked so much that they set papers on fire and the lines had to be attached to the ground to prevent the machines from being irrevocably damaged. In several places, including between Boston and Portland, the electrical currents swooping through the lines from magnetic variations were so great that telegraph operators could work off what they called the "celestial power" alone, after having disengaged the batteries.

The disturbance in the Earth's magnetic field lasted for eleven days.

Not only had the outburst on the surface of the sun penetrated deep into the Earth's atmosphere, it had also coursed through whatever electrical structures humans had made on its surface, rendering them useless. In today's terms, the pulses of electric currents flowing in the magnetosphere and the ionosphere cradled within it (the thick band of atmosphere 75 to 1,000 kilometers above the surface where cosmic and solar rays tear apart atoms, making ions) had produced oscillating magnetic fields at the Earth's surface. By Maxwell's laws, those magnetic fields had caused electrical currents to flow. Their conduits were the Earth's crust and upper mantle. These are known as geomagnetically induced currents, or telluric currents, after the Latin word *tellus* for "earth." The currents were looking for long, easy paths to flow in, driven by the oscillations in the magnetic field. The best ones they found were the grounded, highly conductive lines that humans had set up to move telegraph signals. But the power the currents carried was far too strong for the lines to handle. They were overwhelmed; they overheated and shut down, or sparked. Once the coronal mass ejection ran its course and the changing magnetic fields at the surface of the Earth stabilized, the telluric currents stopped.

Contemporary scientists had already noticed that telegraph lines became disturbed when the auroras were dancing. And a few years earlier, in 1852, Sir Edward Sabine, the architect of the magnetic crusade who obsessively plowed through voluminous global readings from the Magnetic Union of Göttingen to discern patterns, had linked sunspot activity with geomagnetic irregularities on the Earth. That was a surprise, and the first hint that the sun and the Earth's magnetic field might frolic with each other. In fact, Carrington was watching the sun for dark spots on its surface in late August 1859 when he noticed the flare that preceded the superstorm. Nevertheless, many scientists of the day did not believe that the sun's activity could have any effect on terrestrial systems. Even after the Carrington event, many remained unconvinced. Among the champions of the skeptics was Lord Kelvin, who had also stood staunchly behind another incorrect idea: the hard-boiled-egg theory of the Earth's interior.

By the time the second huge superstorm struck, 153 years later, scientists harbored no doubt that the sun was causing the Earth's magnetic field to react. They had been nervously waiting to see when the sun would produce what is now known as a Carrington-class superstorm, and wondering how it would affect the Earth when it hit. It arrived without warning on July 23, 2012, when the sun's magnetic field was in a period of relative calm and no extreme events were expected. The reason why few people have heard of it is that by a quirk of fate, this violent eruption happened on the side of the sun facing away from the Earth. Had it occurred a week earlier, its full force would have been focused on this planet, its inhabitants, and our infrastructure.

Baker has conducted what amounts to a forensic analysis of what happened. The *STEREO-A* (Solar TErrestrial RElations Observatory) spacecraft, positioned in interplanetary space, captured the whole affair, and so did a few other craft nearby. Because they were outside the Earth's magnetosphere, where the interplanetary magnetic field is relatively weak, the blast did not produce currents damaging enough to scotch the spacecraft's equipment, and its instruments recorded information about plasmic speeds throughout the event.

It was far worse than anyone could have imagined and far worse than the Quebec or Halloween event. Again, it began with a solar flare, followed by a coronal mass ejection of unusual speed and strength thrusting a targeted mass of magnetized plasma in fast, swooping clouds out into space. The propulsion of solar energetic particles was among the strongest ever observed. A separate study found that the sun had likely produced a coronal mass ejection in the same region slightly earlier and that its trajectory had plowed a furrow through space, allowing the second one to move at more devastating speeds.

It was at least as strong as the Carrington event. Had it struck the Earth on a day when our planet was in position for the equinox, it would have been about half again as strong as the Carrington event, Baker calculated. Again, no one foresaw this event. The effects on the Earth's electrical infrastructure would have been catastrophic, sending civilization back to a pre-electricity Victorian era, NASA said. People would not have been able to use anything that plugs into the sockets in a wall, for starters. But neither would they have been able to fill cars with gas, use banks, or even flush the toilet, because all those functions, including municipal septic systems, ultimately depend on electricity. The effects would have cascaded through society and the economy, even eventually causing long pipelines to corrode by overwhelming the circuitry that prevents their corrosion. It could have taken years to recover, according to reports analyzing the potential fallout from a superstorm.

Disturbing in its own right, the near miss of 2012 galvanized interest from the US and other governments to predict superstorms in better ways. Geomagnetic disturbances like the near miss have been named a focus of the international electric infrastructure security council, which was set up in 2010 to guard against "black sky hazards." Scientists in the United States have begun making maps setting out risks to the electrical grid from geomagnetic storms. And in October 2015, US president Barack Obama established a detailed national space weather action plan to gather and disseminate more information about the phenomenon.

These analyses of what happened and what would have happened are based on a rare superstorm hitting while the Earth's magnetic field is still strong. But the field has been weakening since before the Carrington event, and is far weaker now than in 1859. What if the poles were reversing and the field was down to one-tenth of its usual strength? Put aside for the moment the risks of rare Carrington-class superstorms. How would normal solar flares, which can happen multiple times a week, affect the Earth? What about the frequent, hard-to-predict coronal mass ejections? Episodes of potentially deadly solar energetic particles? The constant attack of galactic cosmic rays? These are routine occurrences that our magnetic shield protects us from. When the shield is down, what will they do to living creatures? The answers are not pleasant.

CHAPTER 27

lethal patches

Long before geophysicists suspected that a flip of the magnetic poles might be in progress, they began delving into a concept that startled them. It was the early 1960s. The theory of reversals was just beginning to creep into respectability. Its implications were breathtaking. Among them: Did reversals kill off or mutate species and therefore affect patterns of evolution? This suggestion went far beyond the idea that the magnetic field provides a refuge from cosmic radiation and shelters our atmosphere from solar winds that would rip it away. It was metaphysical: Did the inner machinations of the molten core help determine what lives and dies on the crust?

The first salvo, in 1963, stemmed directly from the discovery of the Van Allen belts. What would happen to all that radiation trapped in the belts when the poles reversed? Would solar wind be able to bathe the Earth in radiation, causing rampant genetic mutations? And had it done so during previous pole flips? The author of the page-and-a-half paper, Robert Uffen of the University of Western Ontario, hypothesized: yes.

"It is becoming increasingly apparent that the Earth is a heat engine the internal workings of which have controlled not only geo-logical phenomena such as mountain building, volcanoes, and earth-quakes, but also geochemical phenomena such as the development of the atmosphere and the oceans; geophysical phenomena such as the magnetic field and radiation belts; and even biological phenom-ena like the origin and evolution of life," Uffen concluded.

The next step was to examine the rock record. This time, it wasn't only to look for the magnetic memory locked in rocks, but also at its archive of fossils. It was obvious that reversals did not kill off all life, because life had persisted continuously on the planet for at least 3.6 billion years. But had previous reversals led to mass die-offs? At a first pass, there was little evidence. For one thing, the Earth had ex-perienced just five mass extinctions. But there had been hundreds of reversals and near-reversals. Therefore, reversals didn't cause mass extinctions, or at least not always. The logic of that line of reasoning broke down under scrutiny, though. Reversals last for perhaps a few thousand years—or less—and the paleontological record is rarely precise to that time scale. It's hard even to find a global rock record for such a short period, much less a record of species gone missing forever within it. We have evidence of the five mass extinctions be-cause they spanned millions of years.

As researchers dug further into the data, some peculiarities began to spring up. Two of the mass extinctions coincided with abrupt changes in the tempo of reversals. The first was the one 252 million years ago at the end of the Permian period. It is known as the Great Dying because 95 percent of species on the planet vanished. The sec-ond was the one that killed off the dinosaurs and many other species 65 million years ago at the end of the Cretaceous period. A super-chron, when the Earth's magnetic field did not change for tens of millions of years, came before each. By contrast, during those two mass extinctions, the field reversed many times. One theory was that during the superchrons, species evolved without the need to adjust to the rigors of reversals, and so when reversals came, so did pulses of extinction. That may offer a bit of comfort about the vulnerability of species to a reversal today. Ever since the dinosaurs vanished, we have been in a relatively fast-paced pulse of reversals, which may have built some level of protection into the genetic code of species now on Earth.

By 1971, the scientific exploration had turned to comparing an index of the change in the number of taxonomic animal families over the past 600 million years—a measure of rates of extinction but not mass extinction—against the timing of reversals. There was an astonishingly high correlation, the author, Ian Crain of the Australian National University in Canberra, found. But why? Did reversals foster extinction and, therefore, the emergence of new species to replace them? Pointing to lab experiments, Crain proposed that the low magnetic field itself was the killer, causing difficulty in movement and reproduction.

But perhaps there was another killing mechanism. New findings in the 1970s and 1980s were showing that the magnetic poles are important for navigation in almost every species studied, in both large and surprisingly small ways. Many use the field to find food, mates, breeding spots, and wintering areas. But, for example, radishes also align their roots according to the field and dogs prefer to urinate facing north-south rather than east-west, as long as they are off leash and not in the midst of a geomagnetic storm. What happens when the poles are reversing? Can species that rely on the poles to navigate still get where they need to go? If not, do they die en masse?

What of the perils of radiation? The long-standing belief was that the Earth's thick atmosphere provides a physical barrier against a full blast of solar and cosmic radiation whether the magnetic shield holds or not. Exposure to radiation while you are in an airplane, for example, increases along with altitude and latitude, suggesting that the atmosphere is a filter except near the poles, where field lines converge. But what if the field were decimated? A clue came from ocean sediment records. They showed an increase in radioactive beryllium, a marker of the collision of cosmic particles with the atmosphere during the last reversal. That meant more cosmic particles were getting into the upper atmosphere before colliding and scattering damaging secondary radiation. But it was not a sign that the destructive energetic particles themselves were reaching the surface, just that secondary radiation was.

And then there was the investigation into damage not from ionizing particles but from a lack of ozone. The Dutch chemist Paul Crutzen, who won a Nobel Prize in 1995 for his work on the ozone hole, showed in 1975 that when solar protons produced ions in the stratosphere, that led, through other chemical reactions, to the widespread destruction of the ozone layer. In turn, that allowed damaging ultraviolet radiation to reach the surface of the Earth. Other investigators found that during a reversal, vast swaths of ozone would vanish, allowing greater amounts of ultraviolet B radiation to strike the surface of the Earth, especially near whatever poles there were at that time. Ultraviolet B radiation is not ionizing, but it can affect living tissue in myriad destructive ways. Skin cancer and long-term damage to the eyes and to the immune system are all linked to the rays. More recently, the French geophysicist Jean-Pierre Valet proposed that the disintegration of the ozone hole could be one factor in the final die-off of the world's Neanderthal population. The last small populations vanished at the time of the Laschamp excursion forty thousand years ago, when the field was at one-tenth of its normal strength. Neanderthals' tendency to be fair-skinned and redheaded suggests they were especially susceptible to ultraviolet B damage, just as modern humans with that coloration are.

In the end, the evidence of how past reversals had affected past life—and therefore how it would affect life during a future reversal—was slender, largely theoretical, and inconclusive. The

German physicists Karl-Heinz Glassmeier and Joachim Vogt, who did an extensive review of the relevant studies in 2010, concluded, "It is yet too early to decide in which way magnetic field driven biochemical effects influence evolution on Earth." The implication is that they do.

I chatted with Baker about these ideas in his expansive office in Boulder, sitting at a large table overlooking the mountains, a wall of books behind us. A Hollywood casting agent would assign him the role of four-star general. He is tall and broad-shouldered with straw-straight, gingery hair. He has learned how to sit stock-still, as if conserving energy for battle. He has testified before the US Congress about the near miss of 2012, and you can see the authority his presence would wield there. You could even conclude that he's solemn, except that from time to time that prairie-dry wit breaks through and a rare smile creeps across his face. He's unusual among scientists. There are those who spend their whole careers looking at a single thing. One might examine coral reefs. Another, the chemistry of plastic polymers. Yet another, the physics of how to create primordial atomic particles. Baker's passion is to put things together across disciplines. He is a synthesizer. There are the leaf inspectors and then there are the forest rangers, is how he puts it. He is the latter. And when he assembles the evidence, he finds a more declarative story about what will happen to life on Earth during a reversal.

Unquestionably, more potentially deadly solar energetic particles will reach closer to the Earth, he said. That access will be episodic rather than continual. In places, these damaging particles will be able to reach the Earth's surface right down to where humans live. It's the same story with galactic cosmic rays, which are a continual threat. The atmosphere will deflect only the slower, less dangerous particles.

The atmosphere will be a double-edged sword when it comes to radiation; it will both protect and harm. As high-energy particles hit the Earth's atmosphere, some will splinter into secondary particles, producing an additional shower of damaging radiation, akin to what the beryllium marker from the last excursion showed. An unanswered question is how well the Earth's atmosphere will withstand the violence of solar wind during the few thousand years or so of a reversal. The general agreement is that a reversal is too short-lived for much atmospheric corrosion to take place. But Baker thinks of Mars. Over time, the relentless solar wind and radiation tore away its atmosphere when that planet's internal magnetic field died. He would like to see scientists do a closer analysis of how the Earth's atmosphere will fare.

He wants to be clear that he does not envision a world with no protection from the terrestrial magnetic field. Instead, the weak multi-pole magnetic field of the reversal will protect parts of the Earth in complex asymmetric bands. They will not follow latitudinal lines. Some mid-latitude portions of the Earth—where humans tend to congregate—will be less protected than others. On the other hand, whole longitudinal lines could be free of any magnetic shield at all. That means there could be radiation hot spots, just as there are ozone hot spots today from holes in the atmosphere's ozone layer. And not just radiation hot spots, but also lethal patches of intense ultraviolet B radiation from an ozone layer chemically abraded by increased upper-atmosphere solar and cosmic radiation.

"To me it's a very real possibility that parts of the planet will not be habitable," he said.

For him, examining the past for clues about the next reversal has its limits. The next reversal will be fundamentally different from any that preceded it for one crucial reason: The world the shield protects today is different from what it was the last time the poles succeeded in reversing 780,000 years ago, or tried to 40,000 years ago. For one thing, there are 7.5 billion humans on it, twice as many as in 1970. Last time the poles reversed, human ancestors were here in small numbers. "It completely changes the game," Baker said.

We have cut down forests, plowed the lands, hunted creatures for meat and sport, changed the chemistry of the air and the ocean through the burning of fossil fuels, built industries and cities and networks of roads. As of 2012, nearly one-third of species that the World Conservation Union had assessed were under threat of extinction. And it's hard for animals to move freely to and find new living spaces not already taken by human civilization and industry. Humans are driving the Earth system, just as geological forces, such as volcanoes, have done in the past. At the same time, the magnetic field, independent of human action and impossible to control, is plotting insurrection. It speaks to Baker of the possibility of a malignant confluence of effects. Of tipping points. Even if in previous times a reversal was not accompanied by widespread destruction, today it might be. When multiple hazards conflate, the result can be unimaginably worse than a single event. What if the magnetic shield is down and a solar storm erupts and there happens to be a giant earthquake?

But in addition to the biological hazards connected to a reversal, there are the dangers to the vast cyber-electric cocoon we have encased ourselves in, stretching from the depths of the ocean into space. It is the central processing system of modern civilization. And particles don't have to reach the surface to do damage. The Earth's atmosphere is populated with satellites, the International Space Station, and airplanes filled with crew and passengers. Solar energetic particles can rip through their sensitive, miniaturized electronics. Telluric currents produced by cosmic plasma–generated magnetic oscillations in the atmosphere can blast the transformers needed for the electrical grid. And satellite timing systems governing the grids could be knocked out, which would unravel the electric and electronic infrastructure. Because the electrical system is so extraordinarily connected, a failure in just one part of it will spread like wildfire across the globe. Never, in the history of the world, has there been this combination of systems that respond so dramatically to the changing magnetic field.

"We are sitting ducks," Baker said.

CHAPTER 28

the cost of catastrophe

The cost of disaster interests the constitutionally cold-blooded insurance industry. Loss of life, limb, and infrastructure is a business consideration. Which is not to say that those in the industry lack compassion. But it is to say that they look at the future through a different lens from most. That's why, for example, some of the earliest people to take a serious look at the impacts on society from increasing carbon dioxide concentrations in the atmosphere were in the insurance industry. They wanted to know what they were up against.

For the same "what if" reasons, the insurance industry is keenly interested in the idea of how magnetic disturbances from space can affect the economy. In fact, since 2015, under British law, insurers must calculate their exposure to the effects of extreme space weather. These are the solar flares, coronal mass ejections, and solar energetic particle episodes that can happen without warning as well as a full-on Carrington-class superstorm. But the actuarial discipline of assessing those costs is in its infancy. It gained impetus after the near miss of 2012. The analyses so far do not extend to the world in the throes of a pole switch with a wasted magnetic field, when storms whose effects outpace the fury of the Carrington will be common. The analysts also confine themselves to geomagnetic disturbances that remain in the upper atmosphere, damaging electrical infrastructure through telluric currents. It's not a vision of what would happen if damaging solar energetic particles hit next door. But if you are looking for clues about the world with a diminished shield, the insurance industry's examinations of solar weather provide a few.

The Helios Solar Storm Scenario, for one. Developed in late 2016 at the Cambridge Centre for Risk Studies, part of Cambridge University's Judge Business School and financed in part by the insurance giant American International Group (AIG), the study is the global insurance industry's first test for exposure to space weather. It is based on interviews with astrophysicists, economists, engineers, utility managers, and catastrophe modelers, among others. It looked at what space weather would cost the US insurance industry, based on three possible scenarios of single-strike damage to American electrical infrastructure. The scenarios ranged from a relatively modest solar storm to a superstorm whose effects could spin out over months. For example, damaged extra-high-voltage transformers used in electrical grids could take a year or more to rebuild and replace.

US insurance industry losses alone were pegged at between $55 billion and nearly $334 billion, depending on how long the damage lasted. Most of that would come from the loss of power to customers. To put that in perspective, the devastation wrought by Hurricane Katrina in 2005 cost the insurance industry $45 billion. The losses from solar weather could beggar some insurance companies, throwing them out of business.

The same Cambridge group took the research further in a study published in 2017, also financially supported by AIG. This one looked at what it would cost the American economy per day if an extreme solar storm hit the electrical grid, and then it added in ripple-down effects on the economies of other nations dependent on trade with the United States, finally arriving at global estimates. The rationale is that the modern economy is so dependent on the electrical grid that failure there would cascade globally.

The study takes as its starting point the fact that most of the activity during an extreme geomagnetic storm happens in the band from 50 to 55 degrees geomagnetic latitude. In the northern hemisphere, that takes in Chicago; Washington, DC; New York; London; Paris; Frankfurt; and Moscow. In the south, it would affect Melbourne and Christchurch. Then it devises four scenarios, each of which reaches across different latitudinal bands in the United States, and therefore into different industries and economic centers. The span of the solar storm would depend on its power. It could move closer to the equator if the storm were severe.

The least expensive scenario affects 8 percent of the US population in a band mainly along the Canadian border. It costs the American economy $6.2 billion a day in direct and indirect losses, or 15 percent of the US daily gross domestic product. Adding in global costs takes the total to $7 billion a day. (All figures are in constant 2011 US dollars.) As the hit spreads across more of the country under other scenarios, costs also grow. The final scenario, affecting all but the most southerly US states and two-thirds of the country's population, comes with a daily price tag of $41.5 billion to the US economy. That makes up 100 percent of daily US economic production. Additionally there would be $7 billion in costs to the global economy, for a total of $48.5 billion a day. In each scenario, every sector of the economy is affected, from manufacturing to finance to mining to construction to government. The countries outside the United States most affected are those whose economies are most intertwined with America's: China, Canada, Mexico, Japan, Germany, and the United Kingdom.

The study's authors take pains to note that they are only looking at the damage to the electric grid in the United States of a one-day event and knock-on effects from that disruption within the United States and other countries. They have not calculated all global costs from a potential solar storm that persisted across time and geography. What if power grids across Asia and Europe fell at the same time? What if, as some other analyses suggest, rounding up the machinery needed to fix the problem could take a decade or more? In other words, this is a limited set of scenarios.

The electrical grid is not the only technology that would be affected by a powerful solar storm. In a study financed by the UK Space Agency, Jonathan Eastwood of the Blackett Laboratory at Imperial College London and others have concluded that the full economic impact of space weather is unknown, even as the number of incidents is likely to increase. They call for urgent work to figure out the costs. But in the report, the authors compile a startling list of systems already known to be affected by space weather. Again, they do not look at the world in the throes of a reversal and only look at a magnetic disturbance that remains in the atmosphere with effects on technology at the surface of the Earth.

Like the other reports, this one mentions risks to electrical grids. It also pinpoints risks to highly conductive railway and tram networks from telluric currents and the problem of interrupted service for electrified mass transit. And then there's communications. Satellites can suffer great harm from severe space weather, even though they are built to withstand it. That's why satellites flying over the South Atlantic Anomaly, where there's more radiation in the atmosphere, shut down to protect themselves. Energetic electrons in the outer Van Allen belt produce the equivalent of Leyden jar sparks of static electricity, damaging satellite electronics during a storm. Solar energetic particles can cut through miniaturized components, leaving a trail of damage. The trend today is toward greater miniaturization and the smaller electronics become, the more damage one particle can do.

Mobile telephone systems rely on global navigation satellite system

(GNSS) timing information, which stands to be seriously disrupted by waves in the ionosphere during a severe solar storm, perhaps for days on end. Driverless cars and road charging technologies, both of which are becoming more common, are also dependent on satellite timing systems and they too will be less reliable.

And the number of satellites circling the Earth is poised to grow and become more interconnected, according to a 2017 study on the effects of space weather on the satellite industry. New uses for satellites, including for Internet and taking images, is prompting the industry, already worth $208 billion in 2015, to plan for large new fleets of small craft. As of 2017, for example, Boeing was working on plans for a fleet of thousands and SpaceX for a fleet of 4,425. Some of the newest rely on communications between satellites, meaning if one goes down, it affects many. But many of these build-up plans have emerged over the past several years of unusually calm solar weather. Not only that, but because satellite industries compete with one another, they don't tend to share information about problems that arise in the environment above the Earth or how to solve them. The study's authors found that the satellite engineers they interviewed were having trouble convincing company owners that it was worth the money to protect satellites from the rigors of space weather. Space, while not quite seen as the benign vacuum of old, was still not appreciated for the malignant creature it can be.

A lesson in the consequences of not understanding the potential of solar storms to interfere with communications systems recently emerged in an account by members of the US Air Force. It was May 1967. Lyndon B. Johnson was president of the United States, and Leonid Brezhnev was general secretary of the Central Committee of the Communist Party of the Soviet Union. The Cold War was in full force, as was the war in Vietnam. The war in the Middle East was a month away. Times were so tense that the US Air Force Strategic Air Command had one-third of its bomber force on alert at all times.

The sun began to erupt with great solar flares, blasting radio and other electromagnetic waves toward the Earth. A coronal mass ejection followed. Those bursts of radio waves interfered with the system set up by the United States and its allies to track ballistic weapons on their way to North America from the Soviet Union. But the military leaders of the day had never seen a solar storm affect their equipment that way and believed that it was deliberate jamming of their system by the Soviets. It was such a politically fraught time that military commanders took the jamming of instruments to be a potential act of war. They were on the point of launching a full-on nuclear aircraft assault when two members of the US government's solar forecasting staff realized what was happening. The attacker was the sun, not the enemy. Had the bombers been launched, military leaders would not have been able to recall them, because communication lines relying on satellites were too scrambled by the magnetic storm. The incident could have changed the course of geopolitical history. How would leaders today handle such an emergency?

If the Earth's poles are switching, then rather than rare bursts of chaos, disrupting events will become commonplace and persistent. Electrical and electronic technologies will become less reliable, more vulnerable to long-term crashes and damage. Unless we devise ways to protect our systems, signs are that we will no longer be able to count on these cornerstones of modern civilization: the electrical infrastructure; the satellites society depends on more and more in ways people barely recognize; our means of communication; how we get around. It's like an intricate structure of dominoes set to fall in unexpected patterns, each failure reinforcing the next.

CHAPTER 29

trout noses and pigeon beaks

Only about a dozen scientists around the world are doing experiments to solve the curious riddle of precisely how living creatures use the Earth's magnetic field to navigate. Michael Winklhofer is one of them. The day I met him, I had taken a flight to Düsseldorf in northern Germany, and then a long cab ride to his office at the University of Duisburg-Essen. He was standing outside the campus's cafeteria building waiting for me when I arrived, hands in his pockets and narrow shoulders hunched against the cold.

This rarefied area of study, known as magnetoreception, was considered crank science in the 1960s. But by the 1980s, a series of careful studies had shown just how common the magnetic sense is among the planet's species, and how crucial to the acts of feeding and reproduction. The discipline took off in prestige. As a student, Winklhofer attended a lecture on magnetotactic bacteria—the ones that respond to the Earth's magnetic field—and was hooked.

Once we were in his office, and my luggage was stowed in a corner, he showed me a short video on his computer of one of the classic experiments. On the screen were bacteria in pond sediments. They need to know up from down because they continually travel between the water and the mud, pivoting between an oxygenated environment and one that doesn't have oxygen. The experiment involves putting a small magnet near the bacteria and then moving it around. The bacteria rotate in perfect harmony with the magnet's movement, acting like the needle on a compass. The bacteria's magnetic sensibility is critical to their survival. Other studies have shown that in parts of the Earth where the magnetic field is parallel to where the mud and water meet—at the magnetic equator, for example—the density of magnetotactic bacteria drops dramatically. Winklhofer played the video several times, immersed in it, still fascinated with what those bacteria were doing. He and others are looking at fossils of magnetotactic bacteria to try to read in them the field's movements over time.

From single-celled bacteria, this magnetic sixth sense spreads upward through the Linnaean taxonomic chart to more complex creatures. It is present in butterflies, honeybees and fruit flies, fish, lobsters, newts and sea turtles, migrating songbirds, whales and wolves, deer, rats, and many other animals. They are born with it, Winklhofer said. Bees make new hives pointing in the same magnetic direction as the hives of their parents. Termite mounds are always aligned north to south. Worms are oriented to the hemisphere they were born in; Australian worms used in lab experiments always point up in test tubes in the northern hemisphere, while North American ones point down. Chinook salmon inherit a magnetic map along with the ability to taste and smell the differences among rivers. Marine turtles can spend decades at sea before returning to the very beach where they were born in order to lay eggs in their turn. Magnetoreception is as intrinsic as the sense of touch. A superb navigational tool, it is unaltered by time of day, season, or weather, present no matter where you are on the planet.

But how do species convert an intangible magnetic field into flesh and blood? There are two leading theories, each of which relies on the presence of unpaired spinning electrons. Both may be in play at the same time, but they're proving hard to pin down, Winklhofer said. Most magnetic species have cells that contain tiny amounts of magnetite, or lodestone, or a related ferrimagnetic substance, whose unpaired spinning electrons can line up to amplify a magnetic force. Deposits of magnetite and other ferrimagnetic molecules have been discovered in living tissue across the web of life. Magnetotactic bacteria, for example, have as much as 2 percent magnetite or a similar substance in their bodies by weight. Mollusks make toothed tongues out of magnetite, and snails that live near hot deep-sea vents make scales like roof tiles out of another ferrimagnetic molecule. Humans have lodestone molecules in our brains, heart, spleen, and liver, although the question of whether we perceive the magnetic force is hotly controversial. One researcher suggested we have consigned it to our subconscious. Perhaps we draw on it now under the cover of magical flashes of insight. The theory is that the magnetite acts as a tiny compass, patching magnetic information into the nervous system, allowing a creature to read the field. It's like having a built-in GPS. Homing pigeons, for example, have six nerve cells containing magnetite at six different spots on the skin of their upper beaks. Rainbow trout have magnetite cells in their noses.

The second idea is that certain molecules containing an unpaired spinning electron each can pair up to become a chemical compass within the body that tracks the field's inclination. The molecules are thought to be contained within a protein and fixed in a cell, perhaps in the retina, so they don't float around. These electrons line up according to the Earth's magnetic field. In birds, they seem to be triggered by the quality of light available. Birds may even be able to process images of the magnetic field in the part of the brain responsible for vision, meaning they can see field lines, or perhaps their absence.

Researchers in magnetoreception grasped early that a pole switch might affect the creatures they studied, both because there could be more than one pair of poles during a reversal and because the field itself would be so weak. Kenneth Lohmann, a biologist at the University of North Carolina and a pioneer in magnetoreception, wrote in a 2008 paper that rapid field changes during a reversal could disrupt on a massive scale animals' ability to return home to nest, leading animals to establish new birthing areas when they can't find their old ones. That's important because it could potentially change survival rates of the young in the short term.

In turn, that's important because some of the species that rely on the magnetic field to orient themselves are endangered today for other reasons than a potential switch in the direction of the Earth's field. Mainly, that means humans have destroyed their habitat or hunted or fished them to the brink. All seven species of migratory sea turtle, for example, are at risk of extinction in the wild. Kemp's ridley sea turtles are in the most danger, according to the World Conservation Union's red list, with a global population of fewer than ten thousand. Many species of whale, which migrate for food and birthing spots, are endangered. One in eight birds is at risk of extinction, and recently the migrating swooping insectivores, such as swallows, have seen sharp declines in numbers across Europe and North America. Bees are in trouble all over the world. Could this be the straw that breaks the camel's back for some species?

Winklhofer said that biologists and geophysicists have been investigating this question for several years and have come to the conclusion that as long as a reversal isn't instantaneous and as long as the species' populations are robust enough to begin with, most will eventually adjust. Researchers working with robins, for example, which are sensitive to inclination, or dip, rather than the north-south direction, have found that they can adjust to a field that has a dramatically lower intensity than they are used to. They can also adjust to poles in radically different places. They just need time. He pointed out that species that navigate using the field are used to recalibrating frequently because the field changes slightly all the time. Songbirds, for example, calibrate their magnetic sensibility at twilight every day, finely attuned to any secular variations.

He pulled up another file on his computer. This one showed a re-creation of the magnetic field at the height of the most recent reversal. Several poles showed up at mid-latitudes. The magnetic equator ran north-south. It looks impossible to navigate, but his analysis is that species may be able to adjust to it, especially those that rely on inclination rather than the north-south direction of the poles. "Biology," he said, "is flexible."

There are two wild cards for the Earth's life forms during a reversal. The first is the depth of the extinction risk already present for species that migrate and navigate using the poles. It's unclear how much the extra stress of a reversal will affect them, just as it's unclear how more routine geomagnetic disturbances do. NASA recently teamed up with two other organizations to investigate whether solar storms, with the accompanying disruption of the Earth's magnetic field, are linked to whale strandings that are common in New Zealand; Australia; and Cape Cod, Massachusetts. The second uncertainty is how much solar and cosmic radiation will strike the Earth's surface during a reversal. Even an extra 5 to 10 percent more radiation will have injurious effects, but scientists have not been able to calculate exactly what it would mean, Winklhofer said: "Without data, this is where science ends."

CHAPTER 30

a suit of stiff black crayon

The sun was quiet during all six Apollo missions that landed humans on the moon. Neither astronauts nor vessels were exposed to violent storms that throw off damaging solar energetic particles. It was an extraordinary piece of luck. But in August 1972, midway between the date the astronauts of Apollo 16 had returned home and when those of Apollo 17 took off on the final voyage, the sun spewed out the biggest storm it had produced in a century.

Earlier in his career, as Daniel Baker became more involved in NASA and in space weather, he began to wonder what would have happened if a solar storm had struck while astronauts were walking on the moon during that fabled Apollo era. The moon no longer has an internally generated magnetic field or an atmosphere to protect against radiation, leaving astronauts exposed apart from their space suits. He asked what the emergency plan had been. The answer? The astronauts were instructed to dig a hole, and then the more senior would lie down in the hole and the junior would lie on top of him, shielding his superior's body with his own. The hope was that at least one astronaut would be undamaged enough to make it back to

Earth. The August 1972 storm was so strong that any humans exposed to it would have suffered acute radiation sickness and would likely have died, Baker said. "It points out that without the protection of a magnetic field, we are very susceptible."

Space, rather than being the calm, empty, and benign place our forebears envisioned, is full of lethal ionizing radiation. When the poles reverse and the Earth's shield is weakened, some of that solar and galactic radiation will reach into the lower atmosphere and even parts of the surface, Baker said. If people and other species cannot escape to safer parts of the planet, they will suffer the effects of radiation, both debilitating and fatal, some of it akin to what the astronauts would have experienced had they been on the moon in August 1972. During a reversal, Baker expects to see more cancer affecting the eyes, mucous membranes, and stomach lining. He expects to see widespread, acute radiation poisoning of the type seen in the wake of radiation accidents and nuclear warfare. That means both immediate and chronic effects on human health. And while some geophysicists said it's hard to tell how much increased radiation will accompany a reversal and that the fallout may not be as severe as Baker predicted, many said that a common estimate is for cancer rates to increase by 20 percent across the board. That "war on cancer" is looking a lot more challenging.

Scientists and physicians have studied the effects of radiation on living tissue since the short electromagnetic waves, dubbed X-rays—after the scientific unknown x—were discovered in 1895 by Wilhelm Röntgen, the Dutch/German physicist. He famously took an X-ray of the left hand of his wife, Anna, showing the startling, ghostly images of the bones within her hand, plus the outline of the wedding ring on her finger. He won the Nobel Prize in physics for the discovery in 1901. Reports of damage from exposure to the mysterious rays began almost immediately, including burns, hair loss, and death. One of the first deaths from cancer caused by X-ray exposure was

Clarence Dally, a glassblower who worked with the American electricity magnate Thomas Edison in his efforts to make an X-ray focus tube. Being right-handed, Dally repeatedly tested the X-ray on his left hand. When it became too injured, he switched to his right. He died in 1904 at age thirty-nine, but not before his left arm had been amputated at the shoulder, and the right above the elbow in a failed bid to stop the galloping damage. Edison abandoned X-rays in horror.

A year after Röntgen discovered X-rays, the French physicist Henri Becquerel found evidence that uranium spontaneously ejects particles—this is the weak nuclear force at work—making what was soon called a "radioactive" substance. Radioactive materials are also ionizing. They are uncommon in nature. Immediately after Becquerel discovered uranium's odd characteristics, Marie and Pierre Curie experimented with it and discovered the radioactive substances radium and polonium. Together, the three received the Nobel Prize in physics in 1903. As with X-rays, the injuries and deaths from working with spontaneously radioactive material began to mount quickly, although the dangers were not fully recognized for decades. At one time, radium cures were on offer to freshen one's complexion or clear one's bowels. Marie Curie, who carried tubes of radioactive material around in her lab-coat pockets, died in 1936 at age sixty-six from aplastic anemia, or damage to her bone marrow, likely from exposure to radiation. Her notes in the National Library of France in Paris are still encased in lead-lined boxes, a radioactive shield.

Today, most of the work estimating the risks of illness and death from radiation of any sort, whether from radioactive substances or ionizing electromagnetic radiation, comes from research on the survivors of the Hiroshima and Nagasaki atomic bombs dropped in 1945. There is also information about people who were exposed to radiation during nuclear accidents or whose nuclear medicine treatments went awry. When it comes to data on space travelers, the twenty-four

Apollo astronauts are the only humans to have left lower-Earth orbit. In addition, there are records from astronauts who have orbited Earth on shuttles or resided on the International Space Station, all of which activity has taken place within the protection of the Van Allen belts. Any other information on damage from solar and galactic particles is experimental or theoretical.

At the most basic level, radioactive substances and radiation from electromagnetic waves and solar and cosmic energetic particles damage living beings in similar ways. Differences among them have to do with how much energy the particles or waves have, how big the particles are, and how close you are to them. Spontaneous radioactive decay, like the uranium, radium, and polonium the Curies worked with, involves a large, unwieldy atom with too many neutrons that is trying to become stable. A common way for an atom to do that is to throw off subatomic particles, tiny bits of itself. Sometimes a radioactive atom throws off a neutron or two in what's called neutron release. Sometimes it's two neutrons and two protons joined together, making a new helium nucleus with a positive charge. That's called alpha decay. Sometimes it throws off an electron. That's beta decay. Sometimes the protons and neutrons rearrange themselves, like people taking their seats for a concert after a cocktail party, and the atom emits electromagnetic energy in the form of gamma rays. The point for us is that charged particles or energetic neutrons or nuclei or tiny fast electromagnetic waves are being emitted that can cut through cells and damage them.

Take uranium as an example. It exists naturally on Earth in three isotopes. It always has 92 protons in the nucleus—because when the number of protons changes, so does the name of the element—but different numbers of neutrons, either 146, 143, or 142. You add the neutrons to the protons to name the isotope. The most common on Earth is uranium-238, which has 146 neutrons. It stabilizes itself by alpha decay, transforming into thorium-234, with 90 protons and

144 neutrons, shedding a helium nucleus (two protons, two neutrons) in the process. The isotopes created during radioactive decay are called daughters. Eventually, after many alchemical transformations, uranium-238 becomes lead-206—boring and stable.

Some radioactive isotopes lend themselves to fission, meaning you can bombard them with neutrons to prompt them to split into lighter isotopes that then spontaneously keep splitting and releasing energy in a chain reaction. Uranium-235, with 143 neutrons, is prone to chain reactions. Uranium-238 can be converted into plutonium-239, which chain reacts. The bomb that struck Hiroshima contained uranium-235; the one dropped on Nagasaki, plutonium-239. Scientists have also learned how to harness the power of these types of radioactive chain reactions in nuclear power reactors to produce electricity, often using uranium-235.

Here's how it connects to a reversal. All living things are made from atoms bonded into molecules through their electrons. Ionizing radiation and radioactive emissions break the bonds, either damaging the cell directly or creating knock-on chemical changes in a cell that can lead to its damage or death. As they break the bonds, they free up electrons, setting them in motion and endowing them with enough energy to ionize and excite other molecules in the tissue along a track of damage known as a linear energy transfer, or LET. The strength of the energy transferred is measured in megaelectron volts, or MeVs, named after Alessandro Volta, who invented the voltaic pile. X-rays are low LET. Galactic cosmic rays are very high. The energy transfer can create highly unstable ions within a tissue. The ions want to gain stability, so they scavenge bits from other molecules, damaging tissue in the process. A prime spot for an unstable ion to grab something is from a strand of DNA. Medical articles on the damage from ionizing radiation often feature microscope photographs of DNA. The tracks left by the radiation resemble rips left by a jagged knife dragged through a ribbon.

Astronauts are considered radiation workers, just like people who work with nuclear reactors. Their main occupational hazard has been pegged to be the risk of cancer from radiation. Their cumulative exposure over time is carefully tracked, and they wear dosimeters to record radiation during their missions. But radiation is tricky. The damage from a long, slow exposure of a fixed amount would be different from a short, intense exposure of that same amount. Chronic health problems, leading potentially to cancer, are linked to the long, slow exposure. The prime space risk for slow exposure comes from galactic cosmic rays. Those rays, which are high-energy protons and nuclei thought to have been created by explosions of supernovas in the Milky Way, are thrust through space with so much force that some of them are unstoppable by any shield. They carry far more energy than even the most powerful solar particles. They may be even more apt to cause the biological injuries that lead to cancer than other types of radiation, for unknown reasons, NASA has said.

By contrast, the prime risk for catastrophic immediate injury, called acute radiation sickness, is from a large solar particle event. That involves quick, devastating exposure that overwhelms the body's threshold for radiation. The 200,000 or so people who died in the atomic bombings in Japan in 1945 were killed by acute poisoning, and so was Alexander Litvinenko, a former Russian officer who died in London in 2006 after being slipped the radioactive polonium-210, possibly in a cup of tea. It took him three weeks to die. Images of him, hairless and nearly without eyelashes, lying sick in a hospital bed, flooded the media. Acute radiation sickness begins swiftly. Cells that reproduce quickly in the body, such as hair follicles, those lining the gut, and blood-making cells in bone marrow, are the first to be affected. Nausea comes. Then vomiting. Fatigue, anorexia, and fever follow, and then hemorrhage as bone marrow fails. Death follows if the bone marrow is badly damaged enough. A bone marrow transplant can occasionally save a life if the exposure was relatively low.

But while cancer is the most feared outcome and is perceived as the highest risk, exposure to high levels of radiation has been found to carry many other health effects. Impaired immunity leading to bacterial and viral illness, short-term memory loss, increased risk of heart attack, and blindness can happen soon after exposure. Over time, the risks are for cataracts, fetal malformations, and sterility. Radiation from galactic cosmic rays, particularly the heavy ions, has recently also been linked to damage to the central nervous system, part of the body that had previously been thought to be able to fend off injury. Now it appears that the radiation can prompt the central nervous system to age long before its time, fostering dementia, Alzheimer's disease, Parkinson's disease, and other forms of cognitive harm in the relatively young.

Driven not by a possible pole switch but by the push to send long-term missions to the unprotected planet of Mars in the 2030s, scientists are trying to collect more information about exactly how space radiation affects living tissue. At the moment, they don't know whether it has precisely the same effects as terrestrial sources of radiation, such as X-rays, gamma rays, and radioactive substances. To test this, and see if they can invent effective shields, they have developed materials that resemble human flesh. Known as tissue-equivalent plastic, the most popular formulation has the appearance of a very stiff black crayon. In 2009, they sent some in a telescope to orbit the moon, carefully calibrated to resemble the thickness of muscle that space radiation would have to penetrate before it reached vulnerable bone marrow. Its job is to measure the amount of energy the particles would deposit in tissue and electronics. Results are still being analyzed.

In the meantime, other researchers sent a radiation detector to Mars along with NASA's *Curiosity* rover, which was on a mission to determine whether the red planet could support any life. The monitor was shielded with the same material intended to protect astronauts who might travel to Mars. But bad news: During the 253 days it took to travel to Mars from Earth, from November 26, 2011, to August 6, 2012, the monitors, even shielded, absorbed so much radiation that enduring it would be tantamount to cutting twenty years off one's life. A separate analysis of the radiation that astronauts would likely encounter on an extended trip to Mars found that a single intense solar energetic particle event could simply kill everyone. So far, there are no effective barriers.

Baker and I had talked for hours by this time, trying to imagine the world of the future.

(Might we have to live underground? I asked. Maybe, he said.) He mused about whether space-weather television channels would become a hot item over time. Quoting, he joked, either Niels Bohr or Yogi Berra, he cracked a rare smile and told me that it's tough to make predictions, especially about the future.

Privately, I took things much further. As the field dwindles, will we become nomads, wandering the Earth with magnetometers to track parts of the planet that have retained remnants of the magnetic field? I wondered if the equinoxes, the two days a year when night and day are the same length, linked to geomagnetic disturbances, will become days of mass terror. Or whether religious sects will emerge to placate an angry sun god, a weird postmodern parallel to the citizens of ancient Egypt and Mexico who worshipped the sun for more benign reasons. Or perhaps necessity will force our magnetic sixth sense to reemerge, allowing us once again, like birds, to see the field in order to survive. I could envision the possibility of cancer communes springing up like the leper colonies of old. Or refuges for the radioactively poisoned or for teenagers whose brains the rays have pushed to early dementia. Will suits of stiff black crayon be all the rage?

And then there were psychological questions. I wondered what it will feel like if the lights, that ultimate symbol of civilization's progress, go out. Or how we know where we are when we have four or eight poles. Or where we come from, once the current north pole moves to the south. How does a species so used to controlling the conditions of life adapt to the fact that this revolution within the core is happening no matter what we do?

I left Baker's office overlooking the Rockies, deep in contemplation about the fate of life as we know it. One thing he had said near the beginning of our time together had fastened itself to my imagination. It was not a fact but an image, told with wonder.

He was talking about the solar dynamo and how, for a time, scientists were sure that they understood it fully. The last solar cycle proved them wrong when their predictions about its activity turned out to be wholly incorrect. That's what he thinks about when he considers the state of knowledge of the Earth's tortured magnetic field. It is weakening. The north pole is on the run. The South Atlantic Anomaly is shifting, gaining ground fast and becoming a stronger agent for change. All these indicate mysterious goings-on below the surface of this spinning magnet we live on. It's as if they are pushing up against an opaque glass and, try as we might, we can make out only their shadowed forms.

notes

PART 1

9. **Richard Feynman, Nobel laureate** Richard Feynman, interview by Christopher Sykes, *Fun to Imagine*, BBC, July 15, 1983.

CHAPTER 1

15. **precise mathematical laws so far to describe reality** Neil Turok, *The Universe Within: From Quantum to Cosmos* (Toronto: Anansi Press, 2012), 46 et passim.

CHAPTER 2

18. **holding electrons in place and allowing atoms to link up into molecules** Sean Carroll, in discussion with the author, December 2016.

18. **a field for each of the fundamental forces and thirteen other fields governing matter** David Tong, "The Real Building Blocks of the Universe," Royal Institution lecture, November 25, 2016, available online at https://www.youtube.com/watch?v=zNVQfWC_evg. As he explains, the thirteen fields have to do with quarks, the electron, neutrinos, and the Higgs.

18. **have a value everywhere in the world** Sean Carroll, in discussion with the author, December 2016.

19. **"to understand invisible angels"** Richard Feynman, *The Feynman Lectures on Physics: Commemorative Issue*, vol. 2 (Pasadena: California Institute of Technology, 1989), 20–29.

19. **a little wave tied up into a bundle of energy** Tong, "The Real Building Blocks."

19. **the bits that will eventually form the cores of atoms** For more detail, consult G. Brent Dalrymple, *Ancient Earth, Ancient Skies: The Age of Earth and Its Cosmic Surroundings* (Stanford: Stanford University Press, 2004).

21. **ascending atomic number, from hydrogen on up** It goes from hydrogen, with one proton, to the handful of man-made atoms with 118, called Oganesson. So far. New elements with even more protons could be yet created in a lab. The bigger they get, the less stable they are, because the strong nuclear force keeping their protons together is struggling to keep up.

21. **Those variations are called isotopes** The nomenclature of an isotope comes from combining its element name and the sum of its protons and neutrons. By far the most common carbon isotope (99 percent) is carbon-12, with six protons and six neutrons. Carbon-13, with six and seven, makes up about 1 percent. Carbon-14, with six and eight, is radioactive and extremely rare and is created when a hot neutron left over from cosmic radiation shows up in the carbon nucleus. This form of carbon is unstable and its mission is to gain greater stability. So, in lockstep with time, one of its neutrons mutates, creating a proton and turning carbon-14 into nitrogen-14. Carbon-14 is the one that scientists use to tell how old a fossil is, a process called radiocarbon dating after the radioactive carbon isotope. They use the transformation of radioactive potassium (K) into argon (Ar) in the same way. That method of dating became a key in proving the theory of pole reversals.

22. **With some exceptions** Plasmas, for example.

22. **must spin in opposite directions** This is Pauli's exclusion principle.

23. **Electrons strongly prefer not to be in pairs** This is known as Hund's rule.

25. **made up of two components a vector** Technically, the Earth's magnetic field is an axial vector. Thanks to Andrew D. Jackson for this note in a communication with the author in December 2016.

CHAPTER 4

36. **unusually strong natural magnets** Vasilios Melfos et al., "The Ancient Greek Names 'Magnesia' and 'Magnetes' and Their Origin from the Magnetite Occurrences at the Mavrovouni Mountain of Thessaly, Central Greece. A Mineralogical-Geochemical Approach," *Archaeological and Anthropological Sciences* 3, no. 2 (2011): 165–72, doi: 10.1007/s12520-010-0048-6.

36. **"into whose embrace iron leaps"** Pliny the Elder, *Natural History* (Loeb Classical Library, 1938), Book 36, 25, doi:10.4159/DLCL.pliny _elder-natural_history.1938.

36. **as the historian A.R.T. Jonkers chronicles** A.R.T. Jonkers, *Earth's Magnetism in the Age of Sail* (Baltimore: Johns Hopkins University Press, 2003), 39–41.

36. **correctly predicted the solar eclipse of May 28, 585 BCE** Joshua J. Mark, "Thales of Miletus," *Ancient History Encyclopedia*, September 2, 2009, http://www.ancient.eu/Thales_of_Miletus/.

37. **wearing bronze slippers** Diogenes Laërtius, "Empedocles, 484–424 B.C.," in *Lives of Eminent Philosophers* 8: 69, available online at http://www.perseus.tufts.edu/hopper/text?doc=Perseus%3Atext%3A1999.01.0258%3Abook%3D8%3Achapter%3D2.

37. **mysterious circular connection** Jonkers, *Earth's Magnetism*, 40.

38. **as a Victorian translator put it** Titus Lucretius Carus, *On the Nature of Things*, trans. Hugh Andrew Johnstone Munro (London: Bell, 1908).

38. **Galileo Galilei, Charles Darwin, and Albert Einstein** Harvard University scholar Stephen Greenblatt tracked the resurrection of Lucretius's work in *The Swerve: How the World Became Modern* (New York: W. W. Norton & Company, 2011).

38. **Gillian Turner, a physicist and historian of magnetism** Gillian Turner, *North Pole, South Pole: The Epic Quest to Solve the Great Mystery of Earth's Magnetism* (New York: The Experiment, 2011), 9–10.

40. **Lucera in Italy, a geopolitically important site** For more about the siege of Lucera, consult Julie Anne Taylor, *Muslims in Medieval Italy: The Colony at Lucera* (New York: Lexington Books, 2005).

43. **Subsequent archeological excavations** John S. Bradford, "The Apulia Expedition: An Interim Report," *Antiquity* 24, no. 94 (June 1950): 84–94.

CHAPTER 5

48. **different from the Earth's geographical pole** Gregory A. Good, "Instrumentation, History of," in *Encyclopedia of Geomagnetism and Paleomagnetism*, eds. David Gubbins and Emilio Herrero-Bervera (Dordrecht, The Netherlands: Springer, 2007), 435 (referred to subsequently as *Encyclopedia of G and P*).

48. **By the early fifteenth century** Jonkers, *Earth's Magnetism in the Age of Sail*, 26.

48. **modern reconstructions** See NOAA's Historical Magnetic Declination map for images: https://maps.ngdc.noaa.gov/viewers/historical _declination/.

49. **Norman measured the dip** Allan Chapman, "Norman, Robert (Flourished 1560–1585)," in *Encyclopedia of G and P*, 707.

49. **A military letter recently uncovered in the Naples state archives** Paolo Gasparini et al., "Macedonio Melloni and the Foundation of the Vesuvius Observatory," in *Journal of Volcanology and Geothermal Research* 53, no. 1–4 (1992), doi:10.1016/0377-0273(92)90070-T.

CHAPTER 6

53. **not celestial but terrestrial** A.R.T. Jonkers, "Geomagnetism, History of," *Encyclopedia of G and P*, eds. David Gubbins and Emilio Herrero-Bervera (Dordrecht, The Netherlands: Springer, 2007), 356–57.

54. **The great conundrum was longitude** For more, read Dava Sobel and William J. H. Andrewes, *The Illustrated Longitude: The True Story of a Lone Genius Who Solved the Greatest Scientific Problem of His Time* (London: Fourth Estate, 1998), and Jonkers, "Geomagnetism."

55. **68 statute miles or 110 kilometers** A.R.T. Jonkers, Andrew Jackson, and Anne Murray, "Four Centuries of Geomagnetic Data from Historical Records," *Review of Geophysics* 41, no. 2 (2003): 2–15, doi: 10.1029/2002rg000115. Or, as Sobel puts it, 60 minutes or 1 degree equals 110 kilometers or 68 statute miles in *Illustrated Longitude*, 7.

55. **spins on an axis at the same rate every day** That was beginning to be commonly understood after the 1543 publication of Nicolaus Copernicus's treatise *De Revolutionibus Orbium Coelestium*, which described a sun-centered solar system, and the Earth spinning on its axis each day.

55. **you have traveled each day** Read Sobel, *Illustrated Longitude*, for the full story.

55. **with a declination angle of 0** Jonkers, "Geomagnetism," 356.

56. **At the time Gilbert began his magnetic research** Stephen Pumfrey, *Latitude and the Magnetic Earth: The True Story of Queen Elizabeth's Most Distinguished Man of Science* (Duxford, Cambridge: Icon Books, 2003), 70.

57. **Back then, it was outrageous** Allan Chapman, "Gilbert, William (1544–1603)," in *Encyclopedia of G and P*, 361.

59. **Gilbert's main aim** Pumfrey, *Latitude and the Magnetic Earth*, 90.

59. **one scholar who has plowed through the later work** Ibid., 91.

60. **To prove his point** Dava Sobel, *Galileo's Daughter: A Historical Memoir of Science, Faith, and Love* (New York: Penguin Books, 2000), 173.

61. **likely as directed by Inquisitional censors** Pumfrey, *Latitude and the Magnetic Earth*, 222.

61. **under strict house arrest** Read Sobel's *Galileo's Daughter* for the full story.

61. **rather than evidence of his fears of official prosecution** Chapman, "Gilbert, William (1544–1603)," *Encyclopedia of G and P*, 361.

61. **Gilbert's colleague William Harvey unhappily discovered** Ibid.

62. **at the heart of the Maker's creation** Jonkers, "Geomagnetism," 357.

CHAPTER 7

65. **one of about ninety volcanoes** L'Équipe Associée de Volcanologie de L'Université de Clermont-Ferrand II, *Volcanologie de la Chaîne des Puys*, 5th ed. (Clermont-Ferrand: Parc naturel régional des Volcans d'Auvergne, 2009), 20.

65. **two thousand pages of Latin to explain how he came to that date** James Barr, "Pre-Scientific Chronology: The Bible and the Origin of the World," *Proceedings of the American Philosophical Society* 143, no. 3 (1999): 379–87, http://www.jstor.org/stable/3181950.

66. **still debated into the late twentieth century** Ronald L. Numbers, "The Most Important Biblical Discovery of Our Time: William Henry Green and the Demise of Ussher's Chronology," *Church History* 69, no. 2 (2000): 257–76, doi:10.2307/3169579.

66. **Both Neptunists and Plutonists visited the Auvergne** *Volcanologie*, 20.

66. **Modern analysis says** Ibid., 144.

66. **the pressure became too great** Ibid., 155.

CHAPTER 8

71. **"plodding, industrious mathematician without a spark of genius"** S.R.C. Malin and Sir Edward Bullard, "The Direction of the Earth's Magnetic Field at London, 1570–1975," *Philosophical Transactions of the Royal Society of London A: Mathematical, Physical and Engineering Sciences* 299, no. 1450 (1981): 357–423, doi:10.1098/rsta.1981.0026.

72. **on a fast canter** Ibid., 359.

72. **Gunter had taken** Ibid., 414.

74. **"A New and Correct CHART"** Edmond Halley, *The Three Voyages of Edmond Halley in the Paramore 1698–1701*, ed. Norman J. W. Thrower (London: Hakluyt Society, 1980), vol. 2.

75. **until the nineteenth century** Julie Wakefield, *Halley's Quest: A Selfless Genius and His Troubled Paramore* (Washington, DC: Joseph Henry Press, 2005), 141.

75. **except when sailing where Halley's curved lines** Ibid.

76. **for a planetary total of four** Sir Alan Cook, "Halley, Edmond (1656–1742)," in *Encyclopedia of G and P*, eds. David Gubbins and Emilio Herrero-Bervera (Dordrecht, The Netherlands: Springer, 2007), 375.

76. **uncannily accurate prediction** Wakefield, *Halley's Quest*, 141.

77. **one unit of intensity** Turner, *North Pole, South Pole*, 106.

77. **an elegant formula** Ibid., 117, gives a full explanation of the maths.

78. **could be shown to be correct** Chris Jones, "Geodynamo," in *Encyclopedia of G and P*, 287.

78. **He enlisted Gauss** Turner, *North Pole, South Pole*, 124.

78. **fourth and most perfect of his mariner's clocks** Sobel and Andrewes, *The Illustrated Longitude*, 132.

79. **Harrison eventually won the reward** Ibid., passim.

80. **near-fanaticism** David Gubbins, "Sabine, Edward (1788–1883)," in *Encyclopedia of G and P*, 891.

80. **"one of the most turbulent periods"** John Cawood, "The Magnetic Crusade: Science and Politics in Early Victorian Britain," *Isis* 70, no. 4 (1979): 493, doi:10.1086/352338.

80. **was newborn** Today it is called the British Science Association.

80. **Sabine already had a passion for magnetism** Gubbins, "Sabine, Edward (1788–1883)," 891.

80. **the scientific mission took on a zeal** Cawood, "The Magnetic Crusade," 517.

80. **the fervor was about proving British scientific supremacy** Ibid., 494.

81. **masterminded the establishment of observatories** Gubbins, "Sabine, Edward (1788–1883)," 891.

81. **Sabine soldiered on** Ibid.

81. **"the greatest scientific undertaking"** William Whewell, quoted by Cawood in "The Magnetic Crusade," 493.

81. **more than thirty permanent observatories** Cawood, "The Magnetic Crusade," 512–13.

82. **complete what Newton had begun** Ibid., 493.
82. **British science historian** Ibid., 516.

CHAPTER 11

101. **"Magnetism and electricity are not independent things"** Feynman, *Lectures on Physics*, 13–16.
101. **electrical field lines can end** Sean Carroll, in discussion with the author, December 2016.
102. **but only when they are moving** Thanks to Sean Carroll for this explanation, in discussion with the author, December 2016.
102. **all magnetism is produced from currents of one sort or another** Feynman, *Lectures*, 13–16.
102. **If you are at rest with respect to an electrical charge** Thanks to Andrew D. Jackson for this explanation in a communication with the author in December 2016.

CHAPTER 12

107. **It was "cosmologically neutral"** John Lewis Heilbron, *Electricity in the 17th and 18th Centuries: A Study of Early Modern Physics* (Berkeley: University of California Press, 1979), 2.
109. **The fellows of the society wrote back** Ibid., 4.
109. **As the modern American historian of science J. L. Heilbron explains** Ibid., 4–5.
110. **"Forty years ago, when one knew nothing about electricity"** Ibid., 6.
110. **as absurd as boxing a light beam inside a soap bubble** Park Benjamin, *The Intellectual Rise in Electricity: A History* (London: Longmans, Green & Co., 1895), 502.
110. **But then in January 1746, the legendary Dutch physicist Pieter van Musschenbroek** His experiments followed the similar independent finding of the Prussian Lutheran cleric Ewald von Kleist a few months earlier. Von Kleist, alas, wrote the descriptions of his experiment so poorly that no one could reproduce them. So the credit for the discovery has gone to van Musschenbroek and is named after his city.
111. **Van Musschenbroek wrote up the experiment in Latin** Benjamin, *The Intellectual Rise in Electricity*, 519.

112. **"I understand nothing and can explain nothing"** Heilbron, *Electricity in the 17th and 18th Centuries*, 314.

112. **Future scientific refinements replaced the jar's water with a lead lining** Patricia Fara, *An Entertainment for Angels: Electricity in the Enlightenment* (Duxford, Cambridge: Icon Books, 2002), 56.

112. **As the Cambridge University historian of science Patricia Fara explains** Ibid., passim.

112. **It was also dangerous** Ibid., 54–55.

113. **they were living in an "age of wonders"** Ibid., 70.

113. **It smacked of a carnival** Ibid., 71.

114. **At one point, he methodically disassembled** Joseph Priestley, *The History and Present State of Electricity: With Original Experiments* (London: printed for C. Bathurst et al., 1775), 201–203, available online at https://archive.org/details/historyandprese00priegoog.

114. **So, on a stormy day in Philadelphia in June 1752** Ibid., 216–20.

115. **The physics of lightning is still being explored today** Joseph R. Dwyer and Martin A. Uman, "The Physics of Lightning," *Physics Reports* 534, no. 4 (2014): 147–241, doi:10.1016/j.physrep.2013.09.004.

115. **searching for a positive charge** Sometimes a negative or positive spark can come up from the ground to meet its opposite in a cloud.

116. **At the British court** Fara, *Entertainment for Angels*, 3.

CHAPTER 13

117. **Together, the two were always on the move** Anja Skaar Jacobsen, "Introduction: Hans Christian Ørsted's Chemical Philosophy," in H. C. Ørsted, *H. C. Ørsted's Theory of Force: An Unpublished Textbook in Dynamical Chemistry,* ed. and trans. Anja Skaar Jacobsen, Andrew D. Jackson, Karen Jelved, and Helge Kragh (Copenhagen: The Royal Danish Academy of Sciences and Letters, 2003), xii.

119. **Ørsted referred to his scientific work as his "literary career"** Andrew D. Jackson and Karen Jelved, "Translators' Note," in *Theory of Force*, xxxiii.

119. **form of religious worship** Andrew D. Wilson, "Introduction," in Hans Christian Ørsted, *Selected Scientific Works of Hans Christian Ørsted*, trans. and ed. Karen Jelved, Andrew D. Jackson, and Ole Knudsen (Princeton: Princeton University Press, 1998), xli.

119. **Because of this overriding philosophy of nature** Robert M. Brain, "Introduction," in Robert M. Brain, Robert S. Cohen, and Ole

Knudsen, eds., *Hans Christian Ørsted and the Romantic Legacy in Science: Ideas, Disciplines, Practices* (Dordrecht, The Netherlands: Springer, 2007), xvi.

122. **Galvani experimented on sheep and frogs, alive and dead** Fara, *An Entertainment for Angels*, 150–52.

124. **who in his 1799 doctoral dissertation** Andrew D. Jackson, in a communication with the author in December 2016.

125. **he would conduct an experiment in class** Wilson, "Introduction," xvii.

129. **It "threatened to upset the whole structure of Newtonian science"** Leslie Pearce Williams, *Michael Faraday: A Biography* (New York: Simon and Schuster, 1971), 140.

129. **a Danish first** Helge Kragh, "Preface," in *H. C. Ørsted's Theory of Force*, ii.

130. **And yet, as the science historian Gerald Holton put it** Gerald Holton, "The Two Maps: Oersted Medal Response at the Joint American Physical Society, American Association of Physics Teachers Meeting, Chicago, January 22, 1980," *American Journal of Physics* 48, no. 12 (1980): 1014–19, doi:10.1119/1.12297.

131. **He spoke for all Britain** Brain, "Introduction," xiv.

CHAPTER 14

134. ***Conversations on Chemistry,* by Jane Marcet** Thanks to Andrew D. Jackson for this note in a communication with the author in December 2016.

134. **tickets to Davy's talks by chance** These and other details are from Williams, *Michael Faraday.*

136. **once described equations as "hieroglyphics"** David Bodanis, *Electric Universe: How Electricity Switched On the Modern World* (New York: Three Rivers Press, 2005), 70.

137. **Faraday did so and tasted fame** Nancy Forbes and Basil Mahon, *Faraday, Maxwell, and the Electromagnetic Field: How Two Men Revolutionized Physics* (Amherst, NY: Prometheus Books, 2014), 61.

137. **The wire moved clockwise around the magnet** This description of Faraday's first electric motor is based on ibid., 59.

138. **"Very satisfactory, but make a more sensible apparatus"** David Gooding, "Nature's School," in David Gooding and Frank A.J.L. James, ed. and introd., *Faraday Rediscovered: Essays on the Life and*

Work of Michael Faraday, 1791–1867 (New York: Stockton Press, 1985), 120.

CHAPTER 15

140. **"A peculiar aura of good nature"** Williams, *Michael Faraday*, 5.
142. **glass furnace installed in his laboratory** Forbes and Mahon, *Faraday, Maxwell, and the Electromagnetic Field*, 63.
142. **fluxes of oxygen in the atmosphere** Frank A.J.L. James, *Michael Faraday: A Very Short Introduction* (Oxford: Oxford University Press, 2010), 83–86.
142. **Ørsted's pioneering ideas were an influence** Thanks to Andrew D. Jackson for this note in a communication with the author in December 2016.
143. **At that point in the history of electromagnetism** Forbes and Mahon, *Faraday, Maxwell, and the Electromagnetic Field*, 69.
144. **On the day he did his experiment** This explanation is based on the description in ibid., 70–73.

CHAPTER 16

147. **This was a far more difficult task** Thanks to Andrew D. Jackson for this note in a communication with the author in December 2016.
148. **"stretched-out or shrunken-down versions of one another"** Turok, *The Universe Within*, 47.
148. **The electromagnetic waves we can see** Ibid.
149. **Maxwell's equations theoretically connected space and time** Ibid.
149. **Neither are time and space separate from each other** Brian Greene, "Introduction," in Albert Einstein, *The Meaning of Relativity: Including the Relativistic Theory of the Non-Symmetric Field* (Princeton: Princeton University Press, 2014), viii–ix.
149. **known as his *annus mirabilis*, or miraculous year** Eleven years later, Einstein followed it with his general theory of relativity, which linked space, time, mass, energy, and gravity.

CHAPTER 17

157. **Its spin helps to organize the field** Thanks to Sabine Stanley for this note in communication with the author in March 2017.
158. **Then something happened** Read Dalrymple, *Ancient Earth, Ancient Skies*, 20–23, for more detail here. Also, thanks to Sabine Stanley of

Johns Hopkins for this and the following explanation in various conversations with the author from 2015 to 2017.

161. **if only we could get at them** Thanks to Sabine Stanley for this in communication with the author in March 2017.

161. **In the case of the sun** Eugene Parker, "Dynamo, Solar," in *Encyclopedia of G and P*, eds. David Gubbins and Emilio Herrero-Bervera (Dordrecht, The Netherlands: Springer, 2007), 178.

162. **including a whole month's worth drawn by Galileo in 1612** Sobel, *Galileo's Daughter*, 58.

CHAPTER 18

165. **Known as the Assam earthquake** Nicolas Ambrasey and Roger Bilham, "Reevaluated Intensities for the Great Assam Earthquake of 12 June 1897, Shillong, India," *Bulletin of the Seismological Society of America* 93, no. 2 (2003): 655–73, doi:10.1785/0120020093.

166. **six competing theories about the structure of the inner Earth** This description is based on Stephen G. Brush, "Chemical History of the Earth's Core," *Eos, Transactions American Geophysical Union* 63, no. 47 (1982): 1185–88, doi:10.1029/EO063i047p01185; Stephen G. Brush, "Nineteenth-Century Debates About the Inside of the Earth: Solid, Liquid or Gas?" *Annals of Science* 36, no. 3 (1979): 225–54, doi:10.1080/000 33797900200231; Stephen G. Brush, "Discovery of the Earth's Core," *American Journal of Physics* 48, no. 9 (1980): 705–24, doi:10.1119/1.12026.

166. **it really came down to a dispute over how old the Earth was** Charles Coulston Gillispie, *Genesis and Geology: A Study in the Relations of Scientific Thought, Natural Theology, and Social Opinion in Great Britain, 1790–1850* (New York: Harper, 1959).

167. **a direct conduit to the seething cauldron below** Brush, "Nineteenth-Century Debates," 228.

167. **an "ejectum from the solar furnace"** Ibid., 229.

167. **"its figure must yield"** Ibid., 239.

168. **The raw egg wobbled a great deal** Ibid., 242.

169. **She was in her teens** Inge Lehmann, "Seismology in the Days of Old," *Eos, Transactions American Geophysical Union* 68, no. 3 (1987): 33–35, doi:10.1029/EO068i003p00033-02.

169. **"the discovery of hell"** Erik Hjortenberg, "Inge Lehmann's Work Materials and Seismological Epistolary Archive," *Annals of Geophysics* 52, no. 6 (2009): 691, doi:10.4401/ag-4625.

170. **gaining entrée into the best society** Andrew D. Jackson, in a conversation with the author in March 2016.

170. **his family only saw him when they ate together** Bruce A. Bolt, "Inge Lehmann: 13 May 1888–21 February 1993," *Biographical Memoirs of Fellows of the Royal Society* 43 (1997): 287, doi:10.1098/rsbm.19997.0016.

170. **woodworking, soccer, and needlepoint** Hjortenberg, "Inge Lehmann's Work Materials," 682.

170. **teacher gave her tougher problems to solve** Bolt, "Inge Lehmann," 287.

170. **Lehmann experienced "severe restrictions"** Ibid., 288.

171. **But it was tolerated** Ibid., 289.

171. **"You should know how many incompetent men"** Ibid., 297.

172. **swap his quiet downscale hotel room** Hjortenberg, "Inge Lehmann's Work Materials," 683.

172. **"Of course I am in the summerhouse"** Ibid., 684.

172. **"a black art"** Bolt, "Inge Lehmann," 291.

172. **cardboard oatmeal boxes** Ibid., 297.

173. **He fobbed her off. For four years** Hjortenberg, "Inge Lehmann's Work Materials," 690–96.

173. **underpins the development of today's theory** David Gubbins, "Lehmann, Inge (1888–1993)," in *Encyclopedia of G and P*, eds. David Gubbins and Emilio Herrero-Bervera (Dordrecht, The Netherlands: Springer, 2007), 469.

173. **chasing keepers in Greenland** Bolt, "Inge Lehmann," 291.

173. **Jeffreys wrote to Bohr** Hjortenberg, "Inge Lehmann's Work Materials," 695.

CHAPTER 19

176. **Japan is a global volcano hot spot** W. Yan, "Japan's Volcanic History, Hidden Under the Sea," *Eos, Transactions American Geophysical Union* 97 (2016), doi:10.1029/2016EO054761.

177. **Few were aligned anywhere in between** This was a miracle, considering the later discovery that the continents had moved.

178. **the American geophysicist Allan Cox** Allan Cox, Richard R. Doell, and G. Brent Dalrymple, "Reversals of the Earth's Magnetic Field," *Science* 144, no. 3626 (1964): 1537–43, doi:10.1126/science.144.3626.1537.

178. **why a material could hold its magnetic charge** Néel also discovered antiferromagnetics, which are substances in whose atoms the spins align in such a way that they fully offset each other. The Néel

point, similar to the Curie point, is the temperature at which an antiferromagnetic loses this alignment.

179. **showed that Brunhes's findings were absolutely correct** Carlo Laj et al., "Brunhes' Research Revisited: Magnetization of Volcanic Flows and Baked Clays," *Eos, Transactions American Geophysical Union* 83, no. 35 (2002): 381–87, doi:10.1029/2002EO000277.

181. **a truck made into a roving rock-sampling lab** Louis Brown, *Centennial History of the Carnegie Institution of Washington: Volume 2, The Department of Terrestrial Magnetism* (Cambridge: Cambridge University Press, 2004), 121.

181. **He turned to Néel** Turner, *North Pole, South Pole*, 173.

181. **"the earth's magnetization has suffered repeated reversals"** J. Hospers, "Summary of Studies on Rock Magnetism," *Journal of Geomagnetism and Geoelectricity* 6, no. 4 (1954): 172–75.

181. **a poll of twenty-eight leading paleomagnetic researchers** Turner, *North Pole, South Pole*, 182–83.

182. **In 1964, Allan Cox, Richard Doell, and Brent Dalrymple** Cox, Doell, and Dalrymple, "Reversals of the Earth's Magnetic Field," 1537–43.

182. **a small tar-paper shack** Konrad Krauskopf, "Allan V. Cox, December 17, 1926–January 27, 1987," in National Academy of Sciences (US), *Biographical Memoirs/National Academy of Sciences of the United States of America* (Columbia University Press; National Academy of Sciences, vol. 71, 1977), 20, https://www.nap.edu/read/5737/chapter/3.

CHAPTER 20

185. **Alfred Wegener gave two public talks** David P. Stern, "A Millennium of Geomagnetism," *Reviews of Geophysics* 40, no. 3 (2002): 17, doi: 10.1029/2000RG000097; Edward Bullard, "The Emergence of Plate Tectonics: A Personal View," *Annual Review of Earth and Planetary Sciences* 3, no. 1 (1975): 3–8, doi:10.1146/annurev.ea.03.050175.000245.

186. **wrote about the backlash** Bullard, "Emergence," 5.

187. **An article in *Time* magazine** Turner, *North Pole, South Pole*, 179.

187. **They re-dubbed the phenomenon "apparent polar wander"** There is something known as "true polar wander." For an explanation, see Vincent Courtillot, "True Polar Wander," in *Encyclopedia of G and P*, eds. David Gubbins and Emilio Herrero-Bervera (Dordrecht, The Netherlands: Springer, 2007), 956–67.

189. **they were underwater volcanoes** Bullard, "Emergence," 10.

189. **dismissed it as "girl talk"** Marie Tharp, "Connect the Dots: Mapping the Seafloor and Discovering the Mid-Ocean Ridge," in *Lamont-Doherty Earth Observatory of Columbia: Twelve Perspectives on the First Fifty Years, 1949–1999*, ed. Laurence Lippsett (New York: Lamont-Doherty Earth Observatory of Columbia, 1999).

189. **amazement, skepticism, and scorn** Ibid.

190. **He became obsessed** Lawrence W. Morley, "Early Work Leading to the Explanation of the Banded Geomagnetic Imprinting of the Ocean Floor," *Eos, Transactions American Geophysical Union* 67, no. 36 (1986): 665–66, doi:10.1029/EO067i036p00665.

190. **the evolution of the ocean basins** Robert S. Dietz, "Continent and Ocean Basin Evolution by Spreading of the Sea Floor," *Nature* 190, no. 4779 (1961): 854–57, doi:10.1038/190854a0.

190. **Morley swiftly wrote up a paper** Morley, "Early Work."

191. **"more appropriately discussed at a cocktail party"** Ibid.

191. **"You don't believe all this rubbish, do you Teddy?"** Bullard, "Emergence," 20.

191. **"I felt cold chills"** Krauskopf, "Allan V. Cox, December 17, 1926–January 27, 1987."

CHAPTER 21

196. **under the supervision of David Gubbins** Gubbins studied under the famous Teddy Bullard.

197. **put together a 380-year record of the field** Jeremy Bloxham and David Gubbins, "The Evolution of the Earth's Magnetic Field," *Scientific American* 261, no. 6 (1989), doi:10.1038/scientificamerican1289-68.

197. **the line ran midway through the Atlantic Ocean** For maps over time, see NOAA's Historical Magnetic Declination map at https://maps.ngdc.noaa.gov/viewers/historical_declination/.

201. **the geophysicist Andrew Jackson** Not the theoretical physicist from the Niels Bohr Institute in Copenhagen who is the expert on Hans Christian Ørsted.

201. **has kept growing, and has kept moving westward** I. Wardinski and R. Holme, "A Time-Dependent Model of the Earth's Magnetic Field and Its Secular Variation for the Period 1980–2000," *Journal of Geophysical Research: Solid Earth* 111, no. B12 (2006): 11, doi:10.1029/2006JB004401.

201. **a massive blob of blue** Ibid.

CHAPTER 22

205. **what the models say the gyre might have looked like in 2015** Christopher C. Finlay, Julien Aubert, and Nicolas Gillet, "Gyre-Driven Decay of the Earth's Magnetic Dipole," *Nature Communications* 7 (2016): 10422, doi:10.1038/ncomms10422.

207. **A paper published in 2016** Javier F. Pavon-Carrasco and Angelo De Santis, "The South Atlantic Anomaly: The Key for a Possible Geomagnetic Reversal," *Frontiers in Earth Science* 4 (2016): 40, doi:10.3389 /feart.2016.00040.

208. **it's not clear that rocks** Jean-Pierre Valet and Alexandre Fournier, "Deciphering Records of Geomagnetic Reversals," *Reviews of Geophysics* 54, no. 2 (2016): 410–46, doi:10.1002/2015RG000506.

209. **A recent paper by the Italian researcher** Leonardo Sagnotti et al., "Extremely Rapid Directional Change During Matuyama-Brunhes Geomagnetic Polarity Reversal," *Geophysical Journal International* 199, no. 2 (2014): 1110–24, doi:10.3389/feart.2016.00040.

209. **it is about twice as strong** Valet and Fournier, "Deciphering Records," passim.

210. **Nonlinear means the answer isn't directly proportional to the sum of the components** Thank you to Sabine Stanley and to Chris Finlay for this explanation, in communications with the author in July 2017.

210. **The most famous way of explaining the idea** Kenneth Chang, "Edward N. Lorenz, a Meteorologist and a Father of Chaos Theory, Dies at 90," *New York Times*, April 17, 2008, http://www.nytimes.com /2008/04/17/us/17lorenz.html.

210. **If a butterfly flaps its wings in Brazil** Edward Lorenz, "The Butterfly Effect," *World Scientific Series on Nonlinear Science Series A 39* (2000): 91–94.

211. **For more than 250 years** June Barrow-Green, *Poincaré and the Three Body Problem* (Providence, RI: American Mathematical Society, 1997), 7.

212. **And here's the sobering truth** Alain Mazaud, "Geomagnetic Polarity Reversals," in *Encyclopedia of G and P*, eds. David Gubbins and Emilio Herrero-Bervera (Dordrecht, The Netherlands: Springer, 2007), 323.

CHAPTER 23

213. **he wrote a famous commentary for *Nature*** Peter Olson, "Geophysics: The Disappearing Dipole," *Nature* 416, no. 6881 (2002): 591–94, doi:10.1038/416591a.

214. **That article accompanied another famous one** Gauthier Hulot et al., "Small-Scale Structure of the Geodynamo Inferred from Oersted and Magsat Satellite Data," *Nature* 416, no. 6881 (2002): 620–23, doi:10.1038/416620a.

214. **"could be the start of a reversal"** David Gubbins, "Earth Science: Geomagnetic Reversals," *Nature* 452, no. 7184 (2008): 165–67, doi:10.1038/452165a.

215. **meticulously outlining the case** Catherine Constable and Monika Korte, "Is Earth's Magnetic Field Reversing?" *Earth and Planetary Science Letters* 246, no. 1 (2006): 1–16, doi:10.1016/j.epsl.2006.03.038.

216. **A French study** Carlo Laj and Catherine Kissel, "An Impending Geomagnetic Transition? Hints from the Past," *Frontiers in Earth Science* 3 (2015): 61, doi:10.3389/feart.2015.00061.

217. **A study by two Italian researchers** Angelo De Santis and Enkelejda Qamili, "Geosystemics: A Systemic View of the Earth's Magnetic Field and the Possibilities for an Imminent Geomagnetic Transition," *Pure and Applied Geophysics* 172, no. 1 (2015): 75–89, doi:10.1007/s00024-014-0912-x.

217. **an ingenious study** John A. Tarduno et al., "Antiquity of the South Atlantic Anomaly and Evidence for Top-Down Control on the Geodynamo," *Nature Communications* 6 (2015), doi:10.1038/ncomms8865.

218. **he stressed the dramatic decay of the dipole** John Tarduno and Vincent Hare, "Does an Anomaly in the Earth's Magnetic Field Portend a Coming Pole Reversal?" *The Conversation*, February 5, 2017, updated February 17, 2017, http://theconversation.com/does-an-anomaly-in-the-earths-magnetic-field-portend-a-coming-pole-reversal-47528.

218. **A fascinating paper published in 2017** Erez Ben-Yosef et al., "Six Centuries of Geomagnetic Intensity Variations Recorded by Royal Judean Stamped Jar Handles," *Proceedings of the National Academy of Sciences* 114, no. 9 (2017): 2160–65, doi:10.1073/pnas.1615797114.

218. **adjured their colleagues to keep heart** Jean-Pierre Valet and Alexandre Fournier, "Deciphering Records of Geomagnetic Reversals," *Reviews of Geophysics* 54, no. 2 (2016): 410–46, doi:10.1002/2015RG000506.

CHAPTER 24

225. **an extensive history of serious sodium fires** Deukkwang An et al., "Suppression of Sodium Fires with Liquid Nitrogen," *Fire Safety Journal* 58 (2013): 204–7, doi:10.1016/j.firesaf.2013.02.001.

226. **no certainty that the dynamo is operating the same way now** Masaru Kono, "Geomagnetism in Perspective," in *Geomagnetism: Treatise on Geophysics,* vol. 5, ed. Masaru Kono (Radarweg, The Netherlands: Elsevier, 2009).

CHAPTER 25

233. **Baker is particularly interested** See Daniel N. Baker and Louis J. Lanzerotti, "Resource Letter SW1: Space Weather," *American Journal of Physics* 84, 166 (2016), doi:10.1119/1.4938403.
235. **the dynamo died** David J. Stevenson, "Dynamos, Planetary and Satellite," *Encyclopedia of G and P,* eds. David Gubbins and Emilio Herrero-Bervera (Dordrecht, The Netherlands: Springer, 2007), 207.
235. **scoured away Mars's atmosphere** "NASA's MAVEN Reveals Most of Mars' Atmosphere Was Lost to Space," NASA Press Release, April 30, 2017, available at https://mars.nasa.gov/news/2017/nasas-maven-reveals-most-of-mars-atmosphere-was-lost-to-space.

CHAPTER 26

240. **clocked at 2,000 kilometers a second** Ramon E. Lopez et al., "Sun Unleashes Halloween Storm," *Eos, Transactions American Geophysical Union* 85, no. 11 (2004): 105–8, doi:10.1029/2004EO110002.
241. **Astronauts at the International Space Station** Donald L. Evans et al., "Service Assessment: Intense Space Weather Storms October 19– November 7, 2003," Silver Spring, MD: NOAA (2004).
242. **More than 100,000 miles of telegraph lines** David H. Boteler, "The Super Storms of August/September 1859 and Their Effects on the Telegraph System," *Advances in Space Research* 38, no. 2 (2006): 159– 72, doi:10.1016/j.asr.2006.01.013.
242. **the benchmark for worst-case, life-threatening exposure** L. W. Townsend et al., "Carrington Flare of 1859 as a Prototypical Worst-Case Solar Energetic Particle Event," *IEEE Transactions on Nuclear Science* 50, no. 6 (2003): 2307–9, doi:10.1109/TNS.2003.821602.
243. **"The light appeared in streams"** Freddy Moreno Cárdenas et al., "The Grand Aurorae Borealis Seen in Colombia in 1859," *Advances in Space Research* 57, no. 1 (2016): 258, doi:10.1016/j.asr.2015.08.026.
243. **"The whole sky appeared mottled red"** Ibid.
243. **whatever horrors the lights foretold** Ibid., passim.

243. **The new telegraph system, with its electrical lines, became a target** Boteler, "Super Storms," 163. The detail on telegraph abnormalities is from his paper, passim.

244. **Carrington was watching the sun for dark spots** Ibid., 160.

244. **Among the champions of the skeptics** Ibid., 170.

245. **A separate study** Ying D. Liu et al., "Observations of an Extreme Storm in Interplanetary Space Caused by Successive Coronal Mass Ejections," *Nature Communications* 5 (2014): 3481, doi:10.1038/ncomms4481.

246. **it would have been about half again as strong as the Carrington event** D. N. Baker et al., "A Major Solar Eruptive Event in July 2012: Defining Extreme Space Weather Scenarios," *Space Weather* 11 (2013): 590, doi:10.1002/swe.20097.

246. **according to reports analyzing the potential fallout** Edward J. Oughton et al., "Quantifying the Daily Economic Impact of Extreme Space Weather Due to Failure in Electricity Transmission Infrastructure," *Space Weather* 15, doi:10.1002/2016SW001491; Mike Hapgood, "Lloyd's 360° Risk Insight Briefing: Space Weather: Its Impact on Earth and Implications for Business," Lloyd's of London, 2010.

CHAPTER 27

249. **The first salvo** Robert J. Uffen, "Influence of the Earth's Core on the Origin and Evolution of Life," *Nature* 198 (1963): 143–44, doi:10.1038/198143b0.

250. **Two of the mass extinctions coincided** J. A. Jacobs, *Reversals of the Earth's Magnetic Field*, 2nd ed. (Cambridge: Cambridge University Press, 1994), 293.

251. **There was an astonishingly high correlation** Ian K. Crain, "Possible Direct Causal Relation Between Geomagnetic Reversals and Biological Extinctions," *Geological Society of America Bulletin* 82 (1971): 2603–6, doi:10.1130/0016-7606(1971)82[2603:PDCRBG]2.0.CO;2.

252. **an increase in radioactive beryllium** G. M. Raisbeck, F. Yiou, and D. Bourles, "Evidence for an Increase in Cosmogenic ^{10}Be During a Geomagnetic Reversal," *Nature* 315 (1985): 315–17, doi:10.1038/315315a0.

252. **widespread destruction of the ozone layer** Karl-Heinz Glassmeier and Joachim Vogt, "Magnetic Polarity Transitions and Biospheric Effects: Historical Perspective and Current Developments," *Space Science Review* 155, no. 1–4 (2010): 400, doi:10.1007/s11214-010-9659-6.

252. **the final die-off of the world's Neanderthal population** Jean-Pierre Valet and Hélène Valladas, "The Laschamp-Mono Lake Geomagnetic Events and the Extinction of Neanderthal: A Causal Link or a Coincidence?" *Quaternary Science Reviews* 29, no. 27–28 (2010): 3887–93, doi:10.1016/j.quascirev.2010.09.010.

253. **The German physicists Karl-Heinz Glassmeier and Joachim Vogt** Glassmeier and Vogt, "Magnetic Polarity Transitions," 406.

CHAPTER 28

258. **Helios Solar Storm Scenario** E. Oughton, J. Copic, A. Skelton, V. Kesaite, Z. Y. Yeo, S. J. Ruffle, M. Tuveson, A. W. Coburn, and D. Ralph, "The Helios Solar Storm Scenario," Cambridge Risk Framework Series, Centre for Risk Studies, University of Cambridge (2016).

258. **in a study published in 2017** E. Oughton et al., "Quantifying the Daily Economic Impact of Extreme Space Weather Due to Failure in Electricity Transmission Infrastructure," *Space Weather* 15, no. 1 (2017): 65–83, doi:10.1002/2016SW001491.

260. **In a study financed by the UK Space Agency** J. P. Eastwood et al., "The Economic Impact of Space Weather: Where Do We Stand?" *Risk Analysis* 37, no. 2 (2017): 206–18, doi:10.1111/risa.12765.

261. **a 2017 study on the effects of space weather on the satellite industry** J. C. Green, J. Likar, and Yuri Shprits, "Impact of Space Weather on the Satellite Industry," *Space Weather* 15, no. 6 (2017): 804–18, doi:10.1002/2017SW001646.

261. **A lesson in the consequences** D. J. Knipp et al., "The May 1967 Great Storm and Radio Disruption Event: Extreme Space Weather and Extraordinary Responses," *Space Weather* 14, no. 9 (2016): 614–33, doi:10.1002/2016SW001423.

CHAPTER 29

263. **his office at the University of Duisburg-Essen** He has since been appointed to a chair at the Institute for Biology and Environmental Studies at the University of Oldenburg in Germany.

264. **There are two leading theories** Michael Winklhofer, "The Physics of Geomagnetic-Field Transduction in Animals," *IEEE Transactions on Magnetics* 45, no. 12 (2009), doi:10.1109/TMAG.2009.2017940.

265. **as much as 2 percent magnetite** Atsuko Kobayashi and Joseph L. Kirschvink, "Magnetoreception and Electromagnetic Field Effects:

Sensory Perception of the Geomagnetic Field in Animals and Humans," in *Electromagnetic Fields Advances in Chemistry* 250 (1995): 368, doi:10.1021/ba-1995-0250.ch021.

265. **we have consigned it to our subconscious** Ibid., 374.

265. **Birds may even be able to process images** Thorsten Ritz et al., "A Model for Photoreceptor-Based Magnetoreception in Birds," *Biophysical Journal* 78, no. 2 (2000): 707–18, doi:10.1016/S0006 -3495(00)76629-X.

266. **could disrupt on a massive scale** Kenneth J. Lohmann et al., "Geomagnetic Imprinting: A Unifying Hypothesis of Long-Distance Natal Homing in Salmon and Sea Turtles," *PNAS* 105, no. 49 (2008): 19096–101, doi:10.1073/pnas.0801859105.

CHAPTER 30

270. **Reports of damage from exposure** K. Sansare et al., "Early Victims of X-Rays," *Dentomaxillofacial Radiology* 40 (2011): 123–25, doi:10.1259 /dmfr/73488299.

271. **He died in 1904 at age thirty-nine** Raymond A. Gagliardi, "Clarence Dally: An American Pioneer," *American Journal of Roentgenology* 157, no. 5 (1991): 922, doi:10.2214/ajr.157.5.1927809.

274. **Astronauts are considered radiation workers** Kira Bacal and Joseph Romano, "Radiation Health and Protection," in *Space Physiology and Medicine: From Evidence to Practice*, eds. Arnaud E. Nicogossian et al. (Dordrecht, The Netherlands: Springer, 2016), 205.

274. **They may be even more apt to cause the biological injuries** Ibid., 214.

275. **high levels of radiation has been found to carry many other health effects** Ibid.

275. **they don't know whether it has precisely the same effects** Jancy McPhee and John Charles, eds., *Human Health and Performance Risks of Space Exploration Missions: Evidence Reviewed by the NASA Human Research Program* (Washington, DC: National Aeronautics and Space Administration, Lyndon B. Johnson Space Center, 2009), 123.

275. **Known as tissue-equivalent plastic** H. E. Spence et al., "CRaTER: The Cosmic Ray Telescope for the Effects of Radiation Experiment on the Lunar Reconnaissance Orbiter Mission," *Space Science Reviews* 150, no. 1 (2010): 243–84, doi:10.1007/s11214-009-9584-8.

275. **Results are still being analyzed** M. D. Looper et al., "The Radiation Environment Near the Lunar Surface: Crater Observations and

Geant4 Simulations," *Space Weather* 11 (2013): 142–52, doi:10.1002/swe .20034.

275. **But bad news** Bacal and Romano, "Radiation Health and Protection," 211.

276. **a single intense solar energetic particle event could simply kill everyone** Susan McKenna-Lawlor et al., "Overview of Energetic Particle Hazards During Prospective Manned Missions to Mars," *Planetary and Space Science* 63–64 (2012): 123–32, doi:10.1016/j.pss.2011.06.017.

selected bibliography

Baker, Daniel N., and Louis J. Lanzerotti. "Resource Letter SW1: Space Weather." *American Journal of Physics* 84, no. 3 (2016): 166–80. doi:10.1119/1.4938403.

Baker, Daniel N., X. Li, A. Pulkkinen, C. M. Ngwira, M. L. Mays, A. B. Galvin, and K.D.C. Simunac. "A Major Solar Eruptive Event in July 2012: Defining Extreme Space Weather Scenarios." *Space Weather* 11, no. 10 (2013): 585–91. doi:10.1002/swe.20097.

Benjamin, Park. *The Intellectual Rise in Electricity: A History.* London: Longmans, Green & Co., 1895.

Bloxham, Jeremy, and David Gubbins. "The Evolution of the Earth's Magnetic Field." *Scientific American* 261, no. 6 (1989): 68–75. doi:10.1038/scientificamerican1289-68.

Bodanis, David. *Electric Universe: How Electricity Switched On the Modern World.* New York: Three Rivers Press, 2005.

Bolt, Bruce A. "Inge Lehmann: 13 May 1888–21 February 1993." *Biographical Memoirs of Fellows of the Royal Society* 43 (1997): 286–301.

Boteler, D. H. "The Super Storms of August/September 1859 and Their Effects on the Telegraph System." *Advances in Space Research* 38, no. 2 (2006): 159–72. doi:10.1016/j.asr.2006.01.013.

Brain, Robert M., Robert S. Cohen, and Ole Knudsen, eds. *Hans Christian Ørsted and the Romantic Legacy in Science: Ideas, Disciplines, Practices.* Dordrecht, The Netherlands: Springer, 2007.

Brush, Stephen G. "Chemical History of the Earth's Core." *Eos, Transactions American Geophysical Union* 63, no. 47 (1982): 1185. doi:10.1029/eo063i047p01185.

———. "Discovery of the Earth's Core." *American Journal of Physics* 48, no. 9 (1980): 705–24. doi:10.1119/1.12026.

———. "Nineteenth-Century Debates About the Inside of the Earth: Solid, Liquid or Gas?" *Annals of Science* 36, no. 3 (1979): 225–54. doi:10.1080/00033797900200231.

Bullard, Edward. "The Emergence of Plate Tectonics: A Personal View." *Annual Review of Earth and Planetary Sciences* 3, no. 1 (1975): 1–31. doi:10.1146/annurev.ea.03.050175.000245.

Cawood, John. "The Magnetic Crusade: Science and Politics in Early Victorian Britain." *Isis* 70, no. 4 (1979): 493–518. doi:10.1086/352338.

Constable, Catherine, and Monika Korte. "Is Earth's Magnetic Field Reversing?" *Earth and Planetary Science Letters* 246, no. 1–2 (2006): 1–16. doi:10.1016/j.epsl.2006.03.038.

Cox, Allan, Richard R. Doell, and G. Brent Dalrymple. "Reversals of the Earth's Magnetic Field." *Science* 144, no. 3626 (1964): 1537–43. doi:10.1126/science.144.3626.1537.

Cárdenas, Freddy Moreno, Sergio Cristancho Sánchez, and Santiago Vargas Dománguez. "The Grand Aurorae Borealis Seen in Colombia in 1859." *Advances in Space Research* 57, no. 1 (2016): 257–67. doi:10.1016/j.asr.2015.08.026.

Dalrymple, G. Brent. *Ancient Earth, Ancient Skies: The Age of Earth and Its Cosmic Surroundings.* Palo Alto: Stanford University Press, 2004.

Dietz, Robert S. "Continent and Ocean Basin Evolution by Spreading of the Sea Floor." *Nature* 190, no. 4779 (1961): 854–57. doi:10.1038/190854a0.

Eastwood, J. P., E. Biffis, M. A. Hapgood, L. Green, M. M. Bisi, R. D. Bentley, R. Wicks, L.-A. Mckinnell, M. Gibbs, and C. Burnett. "The Economic Impact of Space Weather: Where Do We Stand?" *Risk Analysis* 37, no. 2 (2017): 206–18. doi:10.1111/risa.12765.

Einstein, Albert. *The Meaning of Relativity: Including the Relativistic Theory of the Non-Symmetric Field.* Princeton: Princeton University Press, 2014.

Fara, Patricia. *An Entertainment for Angels: Electricity in the Enlightenment.* Cambridge: Icon, 2003.

———. *Fatal Attraction: Magnetic Mysteries of the Enlightenment.* Cambridge: Icon, 2005.

Feynman, Richard Phillips, Matthew Sands, and Robert B. Leighton. *The Feynman Lectures on Physics.* Palo Alto: Stanford University Press, 1989.

Finlay, Christopher C., Julien Aubert, and Nicolas Gillet. "Gyre-Driven Decay of the Earth's Magnetic Dipole." *Nature Communications* 7 (2016): 10422. doi:10.1038/ncomms10422.

Forbes, Nancy, and Basil Mahon. *Faraday, Maxwell, and the Electromagnetic Field: How Two Men Revolutionized Physics.* Amherst, NY: Prometheus Books, 2014.

Glassmeier, Karl-Heinz, and Joachim Vogt. "Magnetic Polarity Transitions and Biospheric Effects." *Space Science Reviews* 155, no. 1–4 (2010): 387–410. doi:10.1007/s11214-010-9659-6.

Gooding, David, and Frank A.J.L. James, eds. *Faraday Rediscovered: Essays on the Life and Work of Michael Faraday, 1791–1867.* New York: Stockton Press, 1985.

Greenblatt, Stephen. *The Swerve: How the World Became Modern.* New York: W. W. Norton & Company, 2011.

Gubbins, David. "Earth Science: Geomagnetic Reversals." *Nature* 452, no. 7184 (2008): 165–67. doi:10.1038/452165a.

Gubbins, David, and Emilio Herrero-Bervera, eds. *Encyclopedia of Geomagnetism and Paleomagnetism.* Dordrecht, The Netherlands: Springer, 2007.

Halley, Edmond. *The Three Voyages of Edmond Halley in the Paramore, 1698–1701.* Edited by Norman J. W. Thrower. Vols. 1, 2. London: Hakluyt Society, 1980.

Heilbron, John Lewis. *Electricity in the 17th and 18th Centuries: A Study of Early Modern Physics.* Berkeley: University of California Press, 1979.

Hjortenberg, Erik. "Inge Lehmann's Work Materials and Seismological Epistolary Archive." *Annals of Geophysics* 52, no. 6 (2009): 679–98. doi:10.4401/ag-4625.

Holton, Gerald. "The Two Maps: Oersted Medal Response at the Joint American Physical Society, American Association of Physics Teachers Meeting, Chicago, 22 January 1980." *American Journal of Physics* 48, no. 12 (1980): 1014–19. doi:10.1119/1.12297.

Jacobsen, Anja Skaar, Andrew D. Jackson, Karen Jelved, and Helge Kragh, eds. *H.C. Ørsted's Theory of Force: An Unpublished Textbook in Dynamical Chemistry.* Copenhagen: Royal Danish Academy of Sciences and Letters, 2003.

James, Frank A.J.L. *Michael Faraday: A Very Short Introduction.* Oxford: Oxford University Press, 2010.

Jelved, Karen, Andrew D. Jackson, and Ole Knudsen, eds. *Selected Scientific Works of Hans Christian Ørsted.* Princeton: Princeton University Press, 1998.

Jonkers, A.R.T. *Earth's Magnetism in the Age of Sail.* Baltimore: Johns Hopkins University Press, 2003.

Jonkers, A.R.T., Andrew Jackson, and Anne Murray. "Four Centuries of Geomagnetic Data from Historical Records." *Reviews of Geophysics* 41, no. 2 (2003): 1006. doi:10.1029/2002rg000115.

Knipp, D. J., A. C. Ramsay, E. D. Beard, A. L. Boright, W. B. Cade, I. M. Hewins, R. H. Mcfadden, W. F. Denig, L. M. Kilcommons, M. A. Shea, and D. F. Smart. "The May 1967 Great Storm and Radio Disruption Event: Extreme Space Weather and Extraordinary Responses." *Space Weather* 14, no. 9 (2016): 614–33. doi:10.1002/2016sw001423.

Kobayashi, Atsuko, and Joseph L. Kirschvink. "Magnetoreception and Electromagnetic Field Effects: Sensory Perception of the Geomagnetic Field in Animals and Humans." *Electromagnetic Fields Advances in Chemistry* 250 (1995): 367–94. doi:10.1021/ba-1995-0250.ch021.

Kono, Masaru, ed. *Geomagnetism: Treatise on Geophysics.* Vol. 5. Radarweg, The Netherlands: Elsevier, 2009.

Laj, Carlo, and Catherine Kissel. "An Impending Geomagnetic Transition? Hints from the Past." *Frontiers in Earth Science* 3, no. 61 (2015). doi:10.3389 /feart.2015.00061.

Laj, Carlo, Catherine Kissel, and Hervé Guillou. "Brunhes' Research Revisited: Magnetization of Volcanic Flows and Baked Clays." *Eos, Transactions American Geophysical Union* 83, no. 35 (2002): 381. doi:10.1029/2002eo000277.

Lehmann, Inge. "Seismology in the Days of Old." *Eos, Transactions American Geophysical Union* 68, no. 3 (1987): 33–35. doi:10.1029/eo068i003p00033-02.

Lippsett, Laurence, ed. *Lamont-Doherty Earth Observatory: Twelve Perspectives on the First Fifty Years, 1949–1999.* Palisades, NY: Lamont-Doherty Earth Observatory of Columbia University, 1999.

Lohmann, Kenneth J., N. F. Putman, and C.M.F. Lohmann. "Geomagnetic Imprinting: A Unifying Hypothesis of Long-Distance Natal Homing in Salmon and Sea Turtles." *Proceedings of the National Academy of Sciences* 105, no. 49 (2008): 19096–101. doi:10.1073/pnas.0801859105.

Malin, S.R.C., and Sir Edward Bullard. "The Direction of the Earth's Magnetic Field at London, 1570–1975." *Philosophical Transactions of the Royal Society A: Mathematical, Physical and Engineering Sciences* 299, no. 1450 (1981): 357–423. doi:10.1098/rsta.1981.0026.

McKenna-Lawlor, Susan, P. Gonçalves, A. Keating, G. Reitz, and D. Matthiä. "Overview of Energetic Particle Hazards During Prospective Manned Missions to Mars." *Planetary and Space Science* 63–64 (2012): 123–32. doi:10.1016 /j.pss.2011.06.017.

Morley, Lawrence W. "Early Work Leading to the Explanation of the Banded Geomagnetic Imprinting of the Ocean Floor." *Eos, Transactions American Geophysical Union* 67, no. 36 (1986): 665. doi:10.1029/eo067i036p00665.

Nicogossian, Arnaud E., R. S. Williams, C. L. Huntoon, C. R. Doarn, J. D. Polk, and V. S. Schneider, eds. *Space Physiology and Medicine: From Evidence to Practice*. Dordrecht, The Netherlands: Springer, 2016.

Oughton, Edward, Jennifer Copic, Andrew Skelton, Viktorija Kesaite, Jaclyn Zhiyi Yeo, Simon J. Ruffle, Michelle Tuveson, Andrew W. Coburn, and Daniel Ralph. "The Helios Solar Storm Scenario." Cambridge Risk Framework Series, Centre for Risk Studies, University of Cambridge, 2016.

Oughton, Edward J., Andrew Skelton, Richard B. Horne, Alan W. P. Thomson, and Charles T. Gaunt. "Quantifying the Daily Economic Impact of Extreme Space Weather Due to Failure in Electricity Transmission Infrastructure." *Space Weather* 15, no. 1 (2017): 65–83. doi:10.1002/2016sw001491.

Priestley, Joseph. *The History and Present State of Electricity, with Original Experiments*. London: Printed for C. Bathurst, and T. Lowndes; J. Rivington, and J. Johnson; S. Crowder, 1775. https://archive.org/details/historyandprese 00priegoog.

Pumfrey, Stephen. *Latitude & the Magnetic Earth*. Duxford, Cambridge: Icon Books, 2003.

Ritz, Thorsten, Salih Adem, and Klaus Schulten. "A Model for Photoreceptor-Based Magnetoreception in Birds." *Biophysical Journal* 78, no. 2 (2000): 707–18. doi:10.1016/s0006-3495(00)76629-x.

Sobel, Dava. *Galileo's Daughter: A Historical Memoir of Science, Faith, and Love*. New York: Penguin Books, 2000.

Sobel, Dava, and William J. H. Andrewes. *The Illustrated Longitude: The True Story of a Lone Genius Who Solved the Greatest Scientific Problem of His Time*. London: Fourth Estate, 1998.

Stern, David P. "A Millennium of Geomagnetism." *Reviews of Geophysics* 40, no. 3 (2002): 1–30 doi:10.1029/2000rg000097.

Tarduno, John A., Michael K. Watkeys, Thomas N. Huffman, Rory D. Cottrell, Eric G. Blackman, Anna Wendt, Cecilia A. Scribner, and Courtney L. Wagner. "Antiquity of the South Atlantic Anomaly and Evidence for Top-Down Control on the Geodynamo." *Nature Communications* 6 (2015): 7865. doi:10.1038 /ncomms8865.

Townsend, L. W., E. N. Zapp, D. I. Stephens, and J. I. Hoff. "Carrington Flare of 1859 as a Prototypical Worst-Case Solar Energetic Particle Event." *IEEE Transactions on Nuclear Science* 50, no. 6 (2003): 2307–309. doi:10.1109/tns.2003.821602.

Turner, Gillian. *North Pole, South Pole: The Epic Quest to Solve the Great Mystery of Earth's Magnetism*. New York: Experiment, 2011.

Turok, Neil. *The Universe Within: From Quantum to Cosmos.* Toronto: Anansi Press, 2012.

Valet, Jean-Pierre, and Alexandre Fournier. "Deciphering Records of Geomagnetic Reversals." *Reviews of Geophysics* 54, no. 2 (2016): 410–46. doi:10.1002/2015rg000506.

Valet, Jean-Pierre, and Hélène Valladas. "The Laschamp-Mono Lake Geomagnetic Events and the Extinction of Neanderthal: A Causal Link or a Coincidence?" *Quaternary Science Reviews* 29, no. 27–28 (2010): 3887–93. doi:10.1016/j.quascirev.2010.09.010.

Wakefield, Julie. *Halley's Quest: A Selfless Genius and His Troubled Paramore.* Washington, DC: Joseph Henry Press, 2005.

Williams, Leslie Pearce. *Michael Faraday: A Biography.* New York: Simon and Schuster, 1971.

Winklhofer, Michael. "The Physics of Geomagnetic-Field Transduction in Animals." *IEEE Transactions on Magnetics* 45, no. 12 (2009): 5259–65. doi:10.1109/tmag.2009.2017940.

acknowledgments

I owe thanks to the many scientists who helped me with this book. You took me from the birth of the universe to the inner turmoil of our own planet and then back out into space. It's been a splendid journey. I stand in awe of the passion each of you brings to the task. And the imagination.

Above all, I want to thank Sabine Stanley of Johns Hopkins University. You were the first person I interviewed for *The Spinning Magnet* (thanks to Harvard's Jerry Mitrovica for the intro), back when it was just a glimmer of an idea. You have been there ever since, with wisdom, patience, and good humor.

The indomitable Jacques Kornprobst, honorary director of l'Observatoire de Physique du Globe de Clermont-Ferrand, not only explained things to me but also drove me around France, introducing me to Bernard Brunhes as well as to the Romanesque wonders of Orcival and Saint-Nectaire. *Merci mille fois.* And the same to Jean-François Lénat of l'Université Blaise Pascal, who encouraged me in the democratization of science this book aspires to.

Andrew D. Jackson of the Niels Bohr Institute in Copenhagen

first guided me through the life of Hans Christian Ørsted and then into the inner workings of atoms and fields. Thank you for helping me understand all that I still needed to learn and for all the work you did on early drafts of the first half of the book.

Conall Mac Niocaill of the University of Oxford spent much of a March afternoon explaining basic concepts of magnetism to me and showing me his experiments. Frank James of the Royal Institute carved time out of a busy day to chat about all things Faraday, including taking me into the archives. Michael Winklhofer spent a full day with me at the University of Duisburg-Essen in Germany to explain how creatures perceive the magnetic field and even bought me lunch and drove me to the train station so I could catch my flight back to London. The enthusiasm of Daniel Lathrop of the University of Maryland, College Park, stays with me still.

Chris Finlay of the Technical University of Denmark has been unfailingly helpful. Thank you so much. And thanks for telling me about the SEDI conference in Nantes, where I had the great good luck to run into Chris Jones, Richard Holme, Cathy Constable, Kathy Whaler, Peter Olson, Christine Thomas, Collin Phillips, Philippe Cardin, Bill McDonough, Hagay Amit, Benoit Langlais, Gauthier Hulot, and others who have influenced this book.

This book would have remained unwritten but for the kindness of Daniel Baker of LASP in Colorado, who both spoke with me at length and met with me to explain his work.

The work of A.R.T. Jonkers and of Gillian Turner has infused the book's research and writing. The magnificent, if now dog-eared, *Encyclopedia of Geomagnetism and Paleomagnetism*, edited by David Gubbins and Emilio Herrero-Bervera, has been my constant companion.

And finally, the theoretical physicist Sean Carroll of Caltech came into my life by chance at the perfect moment to answer burning questions about the four forces of the universe, quantum field theory, and electrons. Thank you for your generosity.

Despite all this help, my mistakes are my own.

I began to imagine this book because of the inquisitiveness, energy, and outright generosity of my agents, Sally Harding and Ron Eckel of the Cooke Agency. Immense thanks to you both.

I have often thought of my editor, Stephen Morrow at Dutton, as the dramaturge of this book, to borrow a theatrical idea. What I mean by that is you alone sensed the shape and scope of this book from the very beginning and gently dropped the exact right ideas at precisely the right times to help it come to life. Your unfailing support, even for the quirky bits, has meant the world to me. Thank you.

Nick Garrison, of Penguin Random House in Canada, took me for a memorable lunch in Toronto near the beginning of the writing of the book and, right there, caused the introduction to be born. Always so devilish to write. Many thanks!

To Madeline Newquist of Dutton, thank you so much for your cheerful patience. To Rachelle Mandik, Dutton's astonishing copy editor, I bow to your gifts.

Many thanks to Nicholas Michel. You are an excellent scientist and a very patient chemistry teacher to your mother.

And to my James, thank you for being my compass, now and always.

index

poles of Earth (*cont.*)
  and Peregrinus' travels, 42
  and Ross expedition, 2
poles of Earth (magnetic)
  and auroras, 242–43
  and Chinese culture, 38–39
  and the core–mantle boundary, 197–99, 201
  and declination change, 71, 75, 212
  and dipole decay, 151, 162–63, 201, 203, 206, 209, 215–19, 227
  and the Earth's rotation, 162
  and geomagnetism, 3, 53–55
  and inclination change, 81
  and longitude problem, 62
  and Peregrinus' compass, 42–43
  and radiation exposure, 241, 246–47, 277
  and Ross expedition, 1–2, 99
  and source of magnetic fields, 24
  and strength of magnetic fields, 206–7
  wandering location of, 187–88, 216
  See also polarity reversals
polonium, 271, 272
Pont Farin (Pontfarein), France, 85–88, 149, 156–57, 179, 183, 216
Priestley, Joseph, 114
Probsthein, Sophie, 98, 124
Proceedings of the Imperial Academy, 176
Prometheus myth, 122
Protestantism, 61, 140
protons, 20–21, 101, 233–35
Puy de Dôme volcano, 29–30, 63–69, 231
P waves, 166, 169, 172, 173

quantum physics, 7, 18, 178
quarks, 19
Quaternary period, 177
Quebec, Canada, 240, 245

radar, 148
radiation
  and description of the Earth's magnetic field, 158
  detectors, 275–76
  exposure to, 251–52, 270, 274
  and gyres in the magnetic field, 162–63
  hot spots, 254
  ionizing, 233, 235, 270, 273

solar, 233, 254, 267, 270
  ultraviolet, 252, 254
radioactivity
  isotopes, 216
  particle emissions, 273
  radioactive beryllium, 252
  radioactive decay, 18, 21, 51, 271, 272–73
radio waves, 148–49, 262
radium, 271, 272
Réaumur, René Antoine Ferchault de, 111–12
relativity, 90, 102–3
remanent magnetism, 179, 190
Rerum Novarum (papal encyclical), 30
"Reversals of the Earth's Magnetic Field" (Cox, Doell, and Dalrymple), 182
reverse-flux patches, 206–7, 214, 217
Reynolds number, 222–23, 226
Ritter, Johann Wilhelm, 118
Roberts, Paul, 224
Roman Catholic Church, 60–61
Roman culture, 41, 42, 64–65
Romanticism, 117
Röntgen, Wilhelm, 270
Ross, James Clark, 1–2, 3, 99
Royal Danish Academy of Sciences and Letters, 128
Royal Institution, 133–36, 139–40, 146, 149, 223, 242
Royal Society of London for Improving Natural Knowledge, 108–9
Russia, 78, 81

Sabine, Edward, 79–83, 197
SAC-C, 200–201
Sagnotti, Leonardo, 209
SAMPEX (Solar Anomalous and Magnetosphere Particle EXplorer), 237
San Andreas fault zone, 192
Sandemanians, 140
satellites and satellite imagery
  and core–mantle boundary, 174
  and core modeling, 162
  and cost of magnetic disturbances, 260
  data and imagery, 91, 162–63, 174, 196
  and evolution of magnetic theory, 150–51
  and global navigation satellite system (GNSS), 260–61

about the author

Alanna Mitchell is an acclaimed science journalist and author of *Sea Sick: The Global Ocean in Crisis*, which won the Grantham Prize for excellence in environmental journalism. She won a National Magazine Award in 2014 for a feature on the biology of extinction, and in 2015 won a New York International Radio Festival Silver Medal for her science documentary on neonicotinoid pesticides. She has contributed to the *New York Times* Science section and CBC Radio's *Quirks & Quarks*.